Basics of Holography

Basics of Holography is an introduction to the subject written by a leading worker in the field.

The first part of the book covers the theory of holographic imaging, the characteristics of the reconstructed image, and the various types of holograms. The practical aspects of holography are then described, including light sources, the characteristics of recording media and practical recording materials, as well as methods for producing different types of holograms and computer-generated holograms. Finally, the most important applications of holography are discussed, such as high-resolution imaging, holographic optical elements, information processing, and holographic interferometry, including vibration analysis and digital electronic techniques. Some useful mathematical results are summarized in appendices, and comprehensive reference sections identify sources of additional information. Numerical problems with their solutions are provided at the end of each chapter.

This book is an invaluable resource for undergraduate and graduate students as well as scientists and engineers who would like to learn more about holography and its applications in science and industry.

PROFESSOR HARIHARAN received his PhD from the University of Kerala in 1958 for his work on photographic resolving power. He became Director of the laboratories at Hindustan Photo Films, Ootacamund, in 1962, and Senior Professor at the Indian Institute of Science, Bangalore, in 1971. In 1973 he joined the Division of Applied Physics of CSIRO in Sydney, retiring as a Chief Research Scientist in 1991. Since retiring, he has continued his research as an Honorary Research Fellow at CSIRO and an Honorary Visiting Professor at the University of Sydney. In 2001 he received the Gold Medal of the International Society for Optical Engineering, in recognition of his outstanding contributions to the field of optics.

Internationally recognized as a leading researcher in holography and interferometry, Professor Hariharan has made outstanding original scientific contributions to several other areas of optics. These include holographic interferometry, laser speckle and speckle interferometry, optical testing, polarization optics and quantum optics. He has published more than 200 papers in international journals as well as four major reviews, and is the author of three books: *Optical Holography* (1984; 2nd edn., 1996), *Optical Interferometry* (1985) and *Basics of Interferometry* (1991).

BASICS OF HOLOGRAPHY

P. HARIHARAN
School of Physics, University of Sydney, Australia

CAMBRIDGE
UNIVERSITY PRESS

PUBLISHED BY THE PRESS SYNDICATE OF THE UNIVERSITY OF CAMBRIDGE
The Pitt Building, Trumpington Street, Cambridge, United Kingdom

CAMBRIDGE UNIVERSITY PRESS
The Edinburgh Building, Cambridge CB2 2RU, UK
40 West 20th Street, New York, NY 10011–4211, USA
477 Williamstown Road, Port Melbourne, VIC 3207, Australia
Ruiz de Alarcón 13, 28014 Madrid, Spain
Dock House, The Waterfront, Cape Town 8001, South Africa

http://www.cambridge.org

First published 2002

Printed in the United Kingdom at the University Press, Cambridge

Typeface Monotype Times NR 11/14 pt   *System* QuarkXPress™   [SE]

A catalogue record for this book is available from the British Library

ISBN 0 521 80741 7 hardback
ISBN 0 521 00200 1 paperback

Contents

Preface

Holography is now used widely as a display medium. In addition, it is firmly established as a tool for scientific and engineering studies, and has found a remarkably wide range of applications for which it is uniquely suited.

This book is intended as an introduction to the subject for science and engineering students, as well as people with a scientific background who would like to learn more about holography and its applications. Key topics are presented at a level that is accessible to anyone with a basic knowledge of physics. A comprehensive bibliography and references to original papers identify sources of additional information. Numerical problems (and solutions) are provided at the end of each chapter, to clarify the principles discussed and give the reader a feel for numbers.

After a brief historical retrospect, the first three chapters review image formation by a hologram, the characteristics of the reconstructed image, and the basic types of holograms, while the next three chapters discuss available light sources, the characteristics of hologram recording media and practical recording materials.

These six chapters are followed by three chapters describing methods for the production of different types of holograms for displays, including multicolor holograms, and methods for making copies of holograms, as well as a chapter describing the production of computer-generated holograms. Following these, the next two chapters review some of the most important technical applications of holography, such as high-resolution imaging, holographic optical elements, and holographic information storage and processing.

Finally, three chapters describe the techniques of holographic interferometry, including the application of digital electronic techniques to holographic interferometry. Some essential mathematical results are summarized in four appendices.

This book had its origin in a series of lectures presented at a Winter School on Optics held at the International Centre for Theoretical Physics in Trieste and organized by the International Commission for Optics, and I would like to express my gratitude for their support.

P. Hariharan
Sydney, November 2001

1
Holographic imaging

A hologram is usually recorded on a photographic plate or a flat piece of film, but produces a three-dimensional image. In addition, making a hologram does not involve recording an image in the conventional sense. To resolve this apparent paradox and understand how holography works, we have to start from first principles.

In conventional imaging techniques, such as photography, what is recorded is merely the intensity distribution in the original scene. As a result, all information about the optical paths to different parts of the scene is lost.

The unique characteristic of holography is the idea of recording both the phase and the amplitude of the light waves from an object. Since all recording materials respond only to the intensity in the image, it is necessary to convert the phase information into variations of intensity. Holography does this by using coherent illumination and introducing, as shown in fig. 1.1, a reference beam derived from the same source. The photographic film records the interference pattern produced by this reference beam and the light waves scattered by the object.

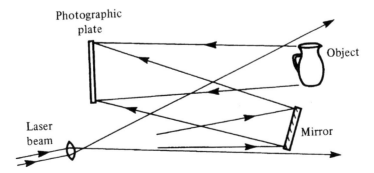

Fig. 1.1. Hologram recording: the interference pattern produced by the reference wave and the object wave is recorded.

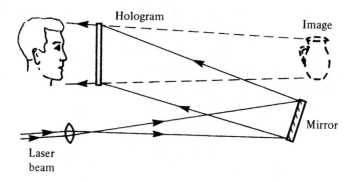

Fig. 1.2. Image reconstruction: light diffracted by the hologram reconstructs the object wave.

Since the intensity at any point in this interference pattern also depends on the phase of the object wave, the resulting recording (the hologram) contains information on the phase as well as the amplitude of the object wave. If the hologram is illuminated once again with the original reference wave, as shown in fig. 1.2, it reconstructs the original object wave.

An observer looking through the hologram sees a perfect three-dimensional image. This image exhibits, as shown in figs. 1.3 and 1.4, all the effects of perspective, and depth of focus when photographed, that characterized the original object.

1.1 Early development

In Gabor's historical demonstration of holographic imaging [Gabor, 1948], a transparency consisting of opaque lines on a clear background was illuminated with a collimated beam of monochromatic light, and the interference pattern produced by the directly transmitted beam (the reference wave) and the light scattered by the lines on the transparency was recorded on a photographic plate. When the hologram (a positive transparency made from this photographic negative) was illuminated with the original collimated beam, it produced two diffracted waves, one reconstructing an image of the object in its original location, and the other, with the same amplitude but the opposite phase, forming a second, *conjugate* image.

A major drawback of this technique was the poor quality of the reconstructed image, because it was degraded by the conjugate image, which was superimposed on it, as well as by scattered light from the directly transmitted beam.

The twin-image problem was finally solved when Leith and Upatnieks

Fig. 1.3. Views from different angles of the image reconstructed by a hologram, showing changes in perspective.

Fig. 1.4. Picture of the reconstructed image taken with the camera lens wide open (f/1.8), showing the effect of limited depth of focus.

[1962, 1963, 1964] developed the off-axis reference beam technique shown schematically in figs. 1.1 and 1.2. They used a separate reference wave incident on the photographic plate at an appreciable angle to the object wave. As a result, when the hologram was illuminated with the original reference beam, the two images were separated by large enough angles from the directly transmitted beam, and from each other, to ensure that they did not overlap.

The development of the off-axis technique, followed by the invention of the laser, which provided a powerful source of coherent light, resulted in a surge of activity in holography that led to several important applications.

1.2 The in-line hologram

We consider the optical system shown in fig. 1.5 in which the object (a transparency containing small opaque details on a clear background) is illuminated by a collimated beam of monochromatic light along an axis normal to the photographic plate.

The light incident on the photographic plate then contains two components. The first is the directly transmitted wave, which is a plane wave whose amplitude and phase do not vary across the photographic plate. Its complex amplitude (see Appendix A) can, therefore, be written as a real constant r. The

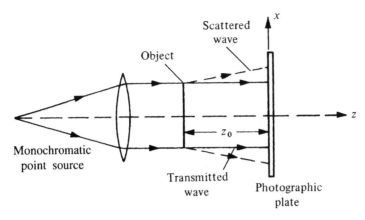

Fig. 1.5. Optical system used to record an in-line hologram.

second is a weak scattered wave whose complex amplitude at any point (x, y) on the photographic plate can be written as $o(x, y)$, where $|o(x, y)| \ll r$.

Since the resultant complex amplitude is the sum of these two complex amplitudes, the intensity at this point is

$$I(x, y) = |r + o(x, y)|^2,$$
$$= r^2 + |o(x, y)|^2 + ro(x, y) + ro^*(x, y), \qquad (1.1)$$

where $o^*(x, y)$ is the complex conjugate of $o(x, y)$.

A 'positive' transparency (the hologram) is then made by contact printing from this recording. If we assume that this transparency is processed so that its amplitude transmittance (the ratio of the transmitted amplitude to that incident on it) can be written as

$$\mathbf{t} = \mathbf{t}_0 + \beta T I, \qquad (1.2)$$

where $\mathbf{t}_0$ is a constant background transmittance, T is the exposure time and β is a parameter determined by the photographic material used and the processing conditions, the amplitude transmittance of the hologram is

$$\mathbf{t}(x, y) = \mathbf{t}_0 + \beta T [r^2 + |o(x, y)|^2 + ro(x, y) + ro^*(x, y)]. \qquad (1.3)$$

Finally, the hologram is illuminated, as shown in fig. 1.6, with the same collimated beam of monochromatic light used to make the original recording. Since the complex amplitude at any point in this beam is, apart from a constant factor, the same as that in the original reference beam, the complex amplitude transmitted by the hologram can be written as

$$u(x, y) = r\mathbf{t}(x, y)$$
$$= r(\mathbf{t}_0 + \beta T r^2) + \beta T r |o(x, y)|^2$$
$$+ \beta T r^2 o(x, y) + \beta T r^2 o^*(x, y). \qquad (1.4)$$

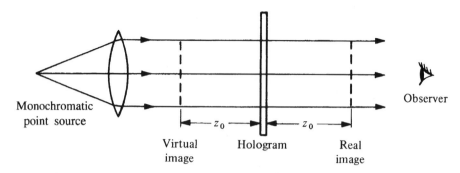

Fig. 1.6.   Optical system used to reconstruct the image with an in-line hologram, showing the formation of the twin images.

The right-hand side of (1.4) contains four terms. The first of these, $r(t_0 + \beta T r^2)$, which represents a uniformly attenuated plane wave, corresponds to the directly transmitted beam.

The second term, $\beta T r \, |o(x, y)|^2$, is extremely small, compared to the other terms, and can be neglected.

The third term, $\beta T r^2 o(x, y)$, is, except for a constant factor, identical with the complex amplitude of the scattered wave from the object and reconstructs an image of the object in its original position. Since this image is formed behind the hologram, and the reconstructed wave appears to diverge from it, it is a virtual image.

The fourth term, $\beta T r^2 o^*(x, y)$, represents a wave similar to the object wave, but with the opposite curvature. This wave converges to form a real image (the conjugate image) at the same distance in front of the hologram.

With an in-line hologram, an observer viewing one image sees it superimposed on the out-of-focus twin image as well as a strong coherent background. Another drawback is that the object must have a high average transmittance for the second term on the right-hand side of (1.4) to be negligible. As a result, it is possible to form images of fine opaque lines on a transparent background, but not *vice versa*. Finally, the hologram must be a 'positive' transparency. If the initial recording is used directly, β in (1.2) is negative, and the reconstructed image resembles a photographic negative of the object.

1.3 Off-axis holograms

To understand the formation of an image by an off-axis hologram, we consider the recording arrangement shown in fig. 1.7, in which (for simplicity) the reference beam is a collimated beam of uniform intensity, derived from the same source as that used to illuminate the object.

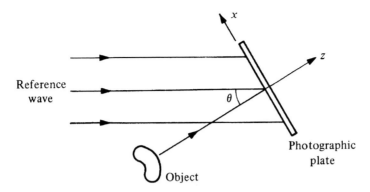

Fig. 1.7.   The off-axis hologram: recording.

The complex amplitude at any point (x, y) on the photographic plate due to the reference beam can then be written (see Appendix A) as

$$r(x, y) = r \exp(i2\pi\xi x), \qquad (1.5)$$

where $\xi = (\sin\theta)/\lambda$, since only the phase of the reference beam varies across the photographic plate, while that due to the object beam, for which both the amplitude and phase vary, can be written as

$$o(x, y) = |o(x, y)| \exp[-i\phi(x, y)]. \qquad (1.6)$$

The resultant intensity is, therefore,

$$
\begin{aligned}
I(x, y) &= |r(x, y) + o(x, y)|^2 \\
&= |r(x, y)|^2 + |o(x, y)|^2 \\
&\quad + r|o(x, y)| \exp[-i\phi(x, y)] \exp(-i2\pi\xi x) \\
&\quad + r|o(x, y)| \exp[i\phi(x, y)] \exp(i2\pi\xi x) \\
&= r^2 + |o(x, y)|^2 + 2r|o(x, y)| \cos[2\pi\xi x + \phi(x, y)]. \qquad (1.7)
\end{aligned}
$$

The amplitude and phase of the object wave are encoded as amplitude and phase modulation, respectively, of a set of interference fringes equivalent to a carrier with a spatial frequency of ξ.

If, as in (1.2), we assume that the amplitude transmittance of the processed photographic plate is a linear function of the intensity, the resultant amplitude transmittance of the hologram is

$$
\begin{aligned}
\mathbf{t}(x, y) = \mathbf{t}_0' &+ \beta\,T\,|o(x, y)|^2 \\
&+ \beta Tr\,|o(x, y)| \exp[-i\phi(x, y)] \exp(-i2\pi\xi x) \\
&+ \beta Tr\,|o(x, y)| \exp[i\phi(x, y)] \exp(i2\pi\xi x), \qquad (1.8)
\end{aligned}
$$

where $\mathbf{t}_0' = \mathbf{t}_0 + \beta Tr^2$ is a constant background transmittance.

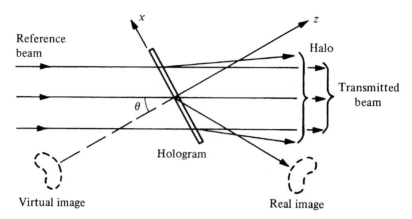

Fig. 1.8.   The off-axis hologram: image reconstruction.

When the hologram is illuminated once again with the original reference beam, as shown in fig. 1.8, the complex amplitude of the transmitted wave can be written as

$$u(x, y) = r(x, y)\mathbf{t}(x, y)$$
$$= \mathbf{t}'_0 r \exp(i2\pi\xi x) + \beta Tr|o(x, y)|^2 \exp(i2\pi\xi x)$$
$$+ \beta Tr^2 o(x, y) + \beta Tr^2 o^*(x, y) \exp(i4\pi\xi x). \tag{1.9}$$

The first term on the right-hand side of (1.9) corresponds to the directly transmitted beam, while the second term yields a halo surrounding it, with approximately twice the angular spread of the object. The third term is identical to the original object wave, except for a constant factor βTr^2, and produces a virtual image of the object in its original position. The fourth term corresponds to the conjugate image which, in this case, is a real image. If the offset angle of the reference beam is made large enough, the virtual image can be separated from the directly transmitted beam and the conjugate image.

In this arrangement, corresponding points on the real and virtual images are located at equal distances from the hologram, but on opposite sides of it. Since the depth of the real image is inverted, it is called a pseudoscopic image, as opposed to the normal, or orthoscopic, virtual image. It should also be noted that the sign of β only affects the phase of the reconstructed image, so that a 'positive' image is always obtained, even if the hologram recording is a photographic negative.

1.4 Fourier holograms

An interesting hologram recording configuration is one in which the complex amplitudes of the waves that interfere at the hologram are the Fourier

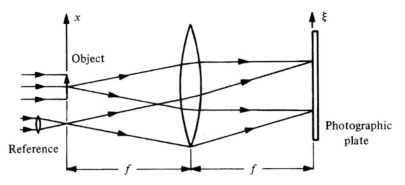

Fig. 1.9. Optical system used to record a Fourier hologram.

transforms (see Appendix B) of the complex amplitudes of the original object and reference waves. Normally, this implies an object of limited thickness, such as a transparency.

To record a Fourier hologram, the object transparency is placed in the front focal plane of a lens, as shown in fig. 1.9, and illuminated with a collimated beam of monochromatic light. The reference beam is derived from a point source also located in the front focal plane of the lens. The hologram is recorded on a photographic plate placed in the back focal plane of the lens [Vander Lugt, 1964].

If the complex amplitude of the wave leaving the object plane is $o(x, y)$, its complex amplitude at the photographic plate located in the back focal plane of the lens is

$$O(\xi,\eta) = \mathcal{F}\{o(x, y)\}. \qquad (1.10)$$

The reference beam is derived from a point source also located in the front focal plane of the lens. If $\delta(x+b, y)$ is the complex amplitude of the wave leaving this point source, the complex amplitude of the reference wave at the photographic plate can be written as

$$R(\xi,\eta) = \exp(-i2\pi\xi b). \qquad (1.11)$$

The intensity in the interference pattern produced by these two waves is, therefore,

$$I(\xi,\eta) = 1 + |O(\xi,\eta)|^2 + O(\xi,\eta)\exp(i2\pi\xi b)$$
$$+ O^*(\xi,\eta)\exp(-i2\pi\xi b). \qquad (1.12)$$

To reconstruct the image, the processed hologram is replaced in the front focal plane of the lens, as shown in fig. 1.10, and illuminated with a collimated beam of monochromatic light.

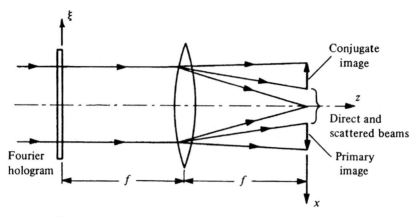

Fig. 1.10.   Image reconstruction by a Fourier hologram.

If the incident wave has unit amplitude, and the amplitude transmittance of the processed hologram is a linear function of the intensity, the complex amplitude of the transmitted wave is

$$U(\xi, \eta) = \mathbf{t}_0 + \beta T I(\xi, \eta). \tag{1.13}$$

The complex amplitude in the back focal plane of the lens is then the Fourier transform of $U(\xi, \eta)$. We have

$$\begin{aligned}
u(x, y) &= \mathcal{F}\{U(\xi, \eta)\} \\
&= (\mathbf{t}_0 + \beta T)\delta(x, y) + \beta T o(x, y) \star o(x, y) \\
&\quad + \beta T o(x - b, y) + \beta T o^* (-x + b, -y).
\end{aligned} \tag{1.14}$$

As shown in fig. 1.10, the wave corresponding to the first term on the right-hand side of (1.14) comes to a focus on the axis, while that corresponding to the second term forms a halo around it. The third term produces an image of the original object, shifted downwards by a distance b, while the fourth term gives rise to a conjugate image, rotated by 180° and shifted upwards by the same distance b.

Fourier holograms have the useful property that the reconstructed image does not move when the hologram is translated in its own plane. This is because a shift of a function in the spatial domain only results in its Fourier transform being multiplied by a phase factor which has no effect on the intensity distribution.

1.5 Lensless Fourier holograms

A hologram with the same properties as a Fourier hologram can be produced, without a lens, with the arrangement shown in fig. 1.11 in which the object is illuminated with a plane wave, and the reference wave comes from a point source in the plane of the object [Stroke, 1965].

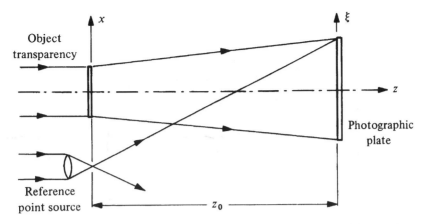

Fig. 1.11.   Optical system used to record a Fourier hologram without a lens.

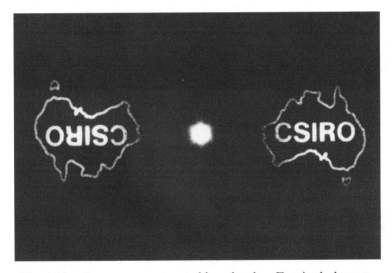

Fig. 1.12.   Images reconstructed by a lensless Fourier hologram.

In this recording arrangement, the effect of the spherical phase factor associated with the near-field (or Fresnel) diffraction pattern of the object transparency (see Appendix C) is eliminated by using a spherical reference wave with the same average curvature.

Figure 1.12 shows the images reconstructed by a lensless Fourier hologram.

1.6 Image holograms

For some applications, there are advantages in recording a hologram of a real image of the object formed by a lens. As shown in fig. 1.13, the hologram plate

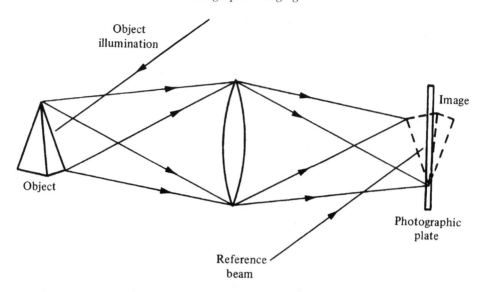

Fig. 1.13.   Recording an image hologram.

is set in the central plane of the image, and a hologram is recorded in the
normal fashion with an off-axis reference beam.

When the hologram is illuminated with the original reference beam, part of
the reconstructed image lies in front of the hologram, and part of the image
lies behind it. Since the image is very close to the hologram plane, the spatial
and temporal coherence requirements for the illumination used to reconstruct
the image are much less critical, and it is even possible, with an object of limited
depth, to use an extended white-light source [Rosen, 1966].

1.7 Reflection holograms

It is also possible to record a hologram with the object beam and the reference
beam incident on the photographic emulsion from opposite sides. The interference fringes then form a series of planes within the emulsion layer, at a small
angle to its surface and about half a wavelength apart. Such holograms, when
illuminated with a point source of white light, reflect a sufficiently narrow band
of wavelengths to reconstruct an image of acceptable quality.

A simpler way of recording such a hologram, with an object of limited
depth, is to use the portion of the reference beam transmitted by the photographic plate to illuminate the object [Denisyuk, 1962].

References

Denisyuk, Yu. N. (1962). Photographic reconstruction of the optical properties of an
    object in its own scattered radiation field. *Soviet Physics – Doklady*, **7**, 543–5.

Gabor, D. (1948). A new microscopic principle. *Nature*, **161**, 777–8.

Leith, E.N. & Upatnieks, J. (1962). Reconstructed wavefronts and communication theory. *Journal of the Optical Society of America*, **52**, 1123–30.

Leith, E. N. & Upatnieks, J. (1963). Wavefront reconstruction with continuous-tone objects. *Journal of the Optical Society of America*, **53**, 1377–81.

Leith, E. N. & Upatnieks, J. (1964). Wavefront reconstruction with diffused illumination and three-dimensional objects. *Journal of the Optical Society of America*, **54**, 1295–301.

Rosen, L. (1966). Focused-image holography with extended sources. *Applied Physics Letters*, **9**, 337–9.

Stroke, G. W. (1965). Lensless Fourier-transform method for optical holography. *Applied Physics Letters*, **6**, 201–3.

Vander Lugt, A. (1964). Signal detection by complex spatial filtering. *IEEE Transactions on Information Theory*, **IT-10**, 139–45.

Problems

1.1. A transmission hologram is recorded using a He–Ne laser ($\lambda = 633$ nm) with the object and reference beams making angles of $+30°$ and $-30°$, respectively, with the normal to the photographic plate. What is the average spatial frequency of the hologram fringes?

The average spatial frequency of the hologram fringes is

$$\xi = \frac{2 \sin 30°}{633 \times 10^{-9}} \, \text{m}^{-1}$$
$$= 1579 \text{ lines}/\text{mm}. \tag{1.15}$$

1.2. What would be the spatial frequency of the fringes in a reflection hologram recorded with a He–Ne laser ($\lambda = 633$ nm) if the object beam is normal to the photographic plate, and the reference beam makes an angle of $45°$ with the normal?

The refractive index of the unexposed photographic emulsion is about 1.6.

As a result, after refraction, the reference beam makes an angle of $26.3°$ with the normal. The angle between the two beams in the photographic emulsion is $153.7°$, and the two beams make angles of $\pm 76.85°$ with the fringe planes, which are inclined at $13.15°$ to the surface of the emulsion.

In addition, the wavelength of the laser in the unexposed photographic emulsion is $633/1.6 = 395.6$ nm. The spatial frequency of the fringes is, therefore,

$$\xi = \frac{2 \sin 76.85°}{395.6 \times 10^{-9}} \, \text{m}^{-1}$$
$$= 4923 \text{ lines}/\text{mm}. \tag{1.16}$$

1.3. A lensless Fourier hologram is recorded with a He–Ne laser ($\lambda = 633$ nm) using an arrangement similar to that shown in fig. 1.11. The object is a transparency with a width of 2 cm placed with its inner edge at a distance of 0.5 cm from the point source providing the reference wave. The hologram is recorded on a photographic plate placed at a distance of 25 cm from the reference source. What is the range of spatial frequencies of the hologram fringes?

At any point in the hologram, the angle between the object and reference beams ranges from a maximum of approximately 0.1 radian (5.73°) to a minimum of approximately 0.02 radian (1.15°). The spatial frequency of the hologram fringes ranges, therefore, from

$$\xi_{max} \approx \frac{\sin 5.73°}{633 \times 10^{-9}} \, m^{-1}$$
$$\approx 158 \text{ lines}/mm \tag{1.17}$$

to

$$\xi_{min} \approx \frac{\sin 1.15°}{633 \times 10^{-9}} \, m^{-1}$$
$$\approx 31.6 \text{ lines}/mm. \tag{1.18}$$

2

The reconstructed image

2.1 Images of a point

To study the characteristics of the reconstructed image and their dependence on the optical system, we consider, as shown in fig. 2.1(a), the hologram of a point object $O(x_O, y_O, z_O)$, recorded with a reference wave from a point source $R(x_R, y_R, z_R)$, using light of wavelength λ_1.

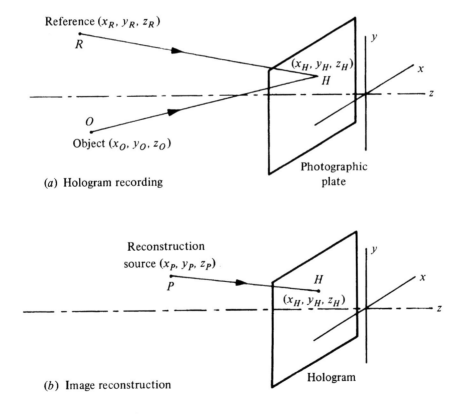

Fig. 2.1. Formation of the image of a point object by a hologram.

15

If, then, this hologram is illuminated with monochromatic light with a wavelength λ_2 from a point source $P(x_P, y_P, z_P)$, as shown in fig. 2.1(b), it can be shown that the coordinates of the primary image of O are [Meier, 1965]

$$x_I = \frac{x_P z_O z_R + \mu x_O z_P z_R - \mu x_R z_P z_O}{z_O z_R + \mu z_P z_R - \mu z_P z_O}, \tag{2.1}$$

$$y_I = \frac{y_P z_O z_R + \mu y_O z_P z_R - \mu y_R z_P z_O}{z_O z_R + \mu z_P z_R - \mu z_P z_O}, \tag{2.2}$$

$$z_I = \frac{z_P z_O z_R}{z_O z_R + \mu z_P z_R - \mu z_P z_O}, \tag{2.3}$$

where $\mu = (\lambda_2 / \lambda_1)$, while those of the conjugate image of O are

$$x_C = \frac{x_P z_O z_R - \mu x_O z_P z_R + \mu x_R z_P z_O}{z_O z_R - \mu z_P z_R + \mu z_P z_O}, \tag{2.4}$$

$$y_C = \frac{y_P z_O z_R - \mu y_O z_P z_R + \mu y_R z_P z_O}{z_O z_R - \mu z_P z_R + \mu z_P z_O}, \tag{2.5}$$

$$z_C = \frac{z_P z_O z_R}{z_O z_R - \mu z_P z_R + \mu z_P z_O}. \tag{2.6}$$

The lateral magnification of the primary image can be defined as

$$M_{\text{lat},I} = dx_I / dx_O = dy_I / dy_O$$

$$= \left[1 + z_O \left(\frac{1}{\mu z_P} - \frac{1}{z_R} \right) \right]^{-1}, \tag{2.7}$$

while that of the conjugate image is

$$M_{\text{lat},C} = \left[1 - z_O \left(\frac{1}{\mu z_P} - \frac{1}{z_R} \right) \right]^{-1}. \tag{2.8}$$

Similarly, the longitudinal magnification of the primary image is

$$M_{\text{long},I} = dz_I / dz_O$$

$$= \frac{1}{\mu} (M_{\text{lat},I})^2, \tag{2.9}$$

while that of the conjugate image is

$$M_{\text{long},C} = -\frac{1}{\mu} (M_{\text{lat},C})^2, \tag{2.10}$$

implying that it has the opposite sign.

If the hologram is illuminated with the reference wave originally used to record it, the primary image has the same size as the original object and coincides with it. However, any change in the position, or the wavelength, of the point source used for reconstruction results in a change in the position and magnification of the reconstructed images and can introduce aberrations.

2.2 Orthoscopic and pseudoscopic images

To understand the implications of the opposite signs of $M_{long,I}$ and $M_{long,C}$, we consider an off-axis hologram recorded with a collimated reference beam incident normal to the photographic plate, as shown in fig. 2.2(a).

When the hologram is illuminated once again with the same collimated reference beam, as shown in fig. 2.2(b), it reconstructs two images, one virtual and the other real. While the virtual image is located behind the hologram, in the same position as the object, the real image is located at the same distance from the hologram, but in front of it. Since corresponding points on the virtual and real images are located at equal distance from the plane of the hologram, the real image has the curious property that its depth appears inverted. Such an image is called a pseudoscopic image.

A hologram that produces an orthoscopic real image of an object can be produced in two steps [Rotz & Friesem, 1966]. In the first step, as shown in fig. 2.3, a hologram (H1) is recorded of the object with a collimated reference beam. H1 is then illuminated once again with the same reference beam, and a second hologram (H2) is recorded of the real image formed by H1 using another collimated reference beam.

When H2 is illuminated with a collimated beam, it reconstructs a pseudoscopic virtual image, located in the same position as the real image formed by H1, and an orthoscopic real image.

A simpler method is to record a hologram of an orthoscopic real image of the object formed by a large concave mirror or lens. When this hologram is illuminated with the original reference beam, it reconstructs the original object wave, producing an orthoscopic real image.

2.3 Image aberrations

If the hologram is replaced in the position in which it was recorded and illuminated with the original reference beam, $z_P = z_R$, and $\mu = 1$. The primary image then coincides with the object. In any other case, the image may exhibit aberrations.

These aberrations can be defined as the phase difference between the reference spheres centered on the image points defined in Section 2.1 and the

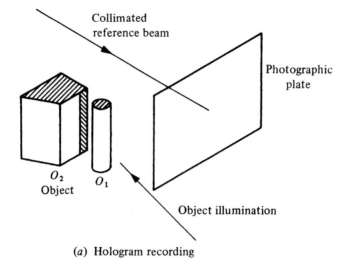

(a) Hologram recording

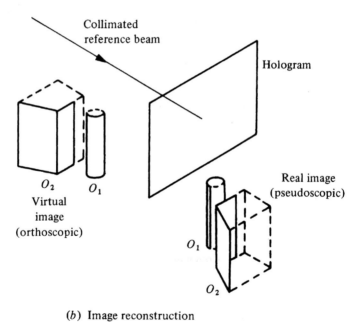

(b) Image reconstruction

Fig. 2.2.   Formation of orthoscopic and pseudoscopic images by a hologram.

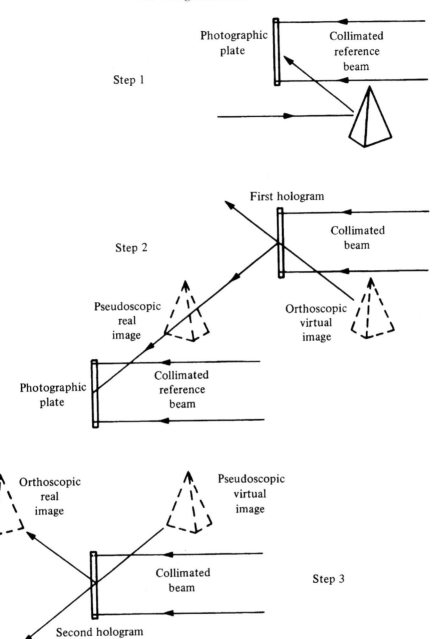

Fig. 2.3. Production of an orthoscopic real image by recording two holograms in succession [Rotz & Friesem, 1966].

actual reconstructed wavefronts. They can be evaluated if we retain the third-degree terms in the expansion for the phase of a spherical wavefront [Meier, 1965].

2.3.1 Classification of aberrations

Hologram aberrations can be classified in the same manner as lens aberrations [Hopkins, 1950]. If we use polar coordinates (ρ, θ) in the hologram plane, the third-order aberration can be written as

$$\Delta\phi_I = (2\pi/\lambda_2) \left[-(1/8) \rho^4 S + (1/2)\rho^3(C_x \cos\theta + C_y \sin\theta) \right.$$
$$- (1/2)\rho^2(A_x \cos^2\theta + A_y \sin^2\theta + 2A_x A_y \cos\theta \sin\theta)$$
$$\left. - (1/4)\rho^2 F + (1/2)\rho(D_x \cos\theta + D_y \sin\theta) \right], \tag{2.11}$$

where S is the coefficient of spherical aberration, C_x, C_y are the coefficients of coma, A_x, A_y are the coefficients of astigmatism, F is the coefficient of field curvature and D_x, D_y are the coefficients of distortion.

For simplicity, we will assume that the object lies on the x axis ($y_O = 0$) and consider only the conjugate (real) image. The aberration coefficients for the primary (virtual) image can be obtained by changing the signs of z_O and z_R in the expressions obtained for the conjugate image.

It can then be shown that the coefficient of spherical aberration is

$$S = (1/z_P^3) - (\mu/z_O^3) + (\mu/z_R^3) - (1/z_C^3). \tag{2.12}$$

When $z_R = z_O$, $z_C = z_P$, and the spherical aberration is zero.

The coefficient of coma is

$$C_x = (x_P/z_P^3) - (\mu x_O/z_O^3) + (\mu x_R/z_R^3) - (x_C/z_C^3). \tag{2.13}$$

Coma can be eliminated only if $z_R = z_O$ and $z_P = \pm z_O$.

The coefficient of astigmatism is

$$A = (x_P^2/z_P^3) - (\mu x_O^2/z_O^3) + (\mu x_R^2/z_R^3) - (x_C^2/z_C^3). \tag{2.14}$$

Astigmatism is eliminated when $z_R = z_O$, $(x_P/z_P) = -(\mu x_R/z_R)$ and $z_P = \mu z_O$. Coma and astigmatism can, therefore, be eliminated simultaneously only when $\mu = 1$.

The coefficient for curvature of field is

$$F = [(x_P^2 + y_P^2)/z_P^3] - [\mu(x_O^2 + y_O^2)/z_O^3]$$
$$+ [\mu(x_R^2 + y_R^2)/z_R^3] - [(x_C^2 + y_C^2)/z_C^3] \tag{2.15}$$

which disappears when the astigmatism is reduced to zero.

Finally, the coefficient of distortion is

$$D_x = [(x_P^3 + x_P y_P^2)/z_P^3] - [\mu(x_O^3 + x_O y_O^2)/z_O^3] \\ + [\mu(x_R^3 + x_R y_R^2)/z_R^3] - [(x_C^3 + x_C y_C^2)/z_C^3]. \qquad (2.16)$$

Distortion cannot normally be eliminated when $\mu \neq 1$.

In particular, unless $M_{lat} = 1$ and $\mu = 1$, the longitudinal magnification is not the same as the lateral magnification, resulting in a distortion in depth. However, longitudinal distortion due to the recording and reconstruction wavelengths not being the same ($\mu \neq 1$) can be minimized by a proper choice of the recording and reconstruction geometry [Hariharan, 1976].

2.4 Image blur

A source other than a laser (one with a finite size and spectral bandwidth) is often used to illuminate the hologram and view the reconstructed image. However, the use of such a source affects the sharpness of the reconstructed image.

From (2.1), it can be shown that if the source used to illuminate the hologram occupies very nearly the same position as the reference source used to record it and has very nearly the same wavelength, the blur of the reconstructed image along the x axis for a source size Δx_P is

$$\Delta x_I = (z_O/z_P)\, \Delta x_P. \qquad (2.17)$$

Similarly, if the source used to illuminate the hologram has a mean wavelength λ_2 very nearly equal to λ_1, the wavelength used to record the hologram, and a spectral bandwidth $\Delta\lambda_2$, the transverse image blur due to the finite spectral bandwidth of the source can be shown to be

$$|\Delta x_I| = (x_P/z_P)z_O(\Delta\lambda_2/\lambda_2). \qquad (2.18)$$

The image blur increases with the depth of the image and the interbeam angle.

It follows from (2.17) and (2.18) that if the central plane of the image lies in the hologram plane, as in an image hologram (see Section 1.6), the restrictions on the size and the spectral bandwidth of the source used to illuminate the hologram can be relaxed. In fact, if the interbeam angle and the depth of the object are small, it is possible to use an extended white-light source to view the image.

2.5 Image speckle

When a diffusely reflecting object is illuminated with light from a laser, each element on its surface produces a diffracted wave. Since the differences in their

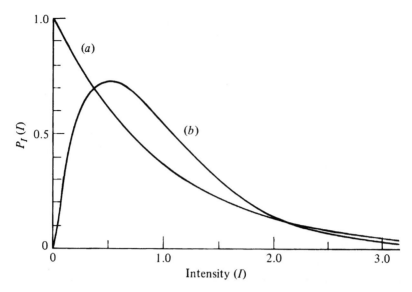

Fig. 2.4.   Intensity distribution in (*a*) a single speckle pattern and (*b*) the incoherent
sum of two different speckle patterns.

optical paths have a random distribution, these diffracted waves interfere with
each other to produce a speckle pattern in the far field (see Appendix D). As a
result, the image reconstructed by a hologram exhibits a speckled appearance.
With polarized light, the intensity in the speckle pattern has, as shown in fig.
2.4(*a*), the negative exponential distribution

$$p(I) = \frac{1}{2\sigma^2} \exp\left(\frac{-I}{2\sigma^2}\right), \tag{2.19}$$

where $\langle I \rangle = 2\sigma^2$ is the mean intensity [Goodman, 1975].

The contrast of the speckle pattern is unity, and its appearance is almost
independent of the nature of the surface, but the size of the speckles increases
with the viewing distance and the *f*-number of the imaging system. With a circular pupil of radius ρ, the average dimensions of the speckles in the image
are

$$\Delta x = \Delta y = 0.61 \lambda f / \rho. \tag{2.20}$$

Speckle is a serious problem in holographic imaging. While several techniques have been described to reduce speckle in the reconstructed image
[McKechnie, 1975], the most common method is to record a number of holograms with the object illuminated from slightly different directions. Each
of these holograms reconstructs the same image, but a different speckle
pattern.

The intensity distribution in the speckle pattern obtained by summing the intensities in two such images with average intensities of $\langle I/2 \rangle$ is, then

$$p(I) = \frac{4I}{\langle I \rangle^2} \exp\left(\frac{-2I}{\langle I \rangle}\right). \tag{2.21}$$

As can be seen from fig. 2.4(*b*), most of the dark areas are eliminated, and the contrast of the pattern is only $1/\sqrt{2}$. If we sum the intensities in the images produced by N holograms, the intensity fluctuations due to speckle are reduced by a factor equal to $\sqrt{N}$.

2.6 Signal-to-noise ratio

Random spatial variations in the intensity of the reconstructed image, commonly caused by scattered light, are referred to as noise. In the case of holographic imaging, when calculating the signal-to-noise ratio, the amplitudes of the signal and the noise must be added, since they are both encoded on a common carrier [Goodman, 1967].

We consider the reconstructed image of a uniform bright patch on a dark background, and assume that the intensity due to the nominally uniform signal is I_S, while that of the randomly varying background is I_N. The noise N in the bright area is given by the variance of the resulting fluctuations of the intensity. It can then be shown that when, as is usually the case, $I_S \gg \langle I_N \rangle$, (where $\langle I_N \rangle$ is the average intensity of the dark background), the signal-to-noise ratio is

$$\frac{I_S}{N} = \left(\frac{I_S}{2\langle I_N \rangle}\right)^{1/2}. \tag{2.22}$$

It follows that even a small amount of scattered light can result in relatively large fluctuations in intensity in the bright areas of the image.

2.7 Image luminance

The luminance of the reconstructed image is directly proportional to the diffraction efficiency of the hologram (defined as the ratio of the energy diffracted into the image to that incident on the hologram) and inversely proportional to the solid angle over which the image can be viewed [Hariharan, 1978].

A hologram recorded of a real image projected either by an optical system or another hologram (see fig. 2.3) also reconstructs an image of the optical system, including any aperture (or pupil) that limits the angular spread of the

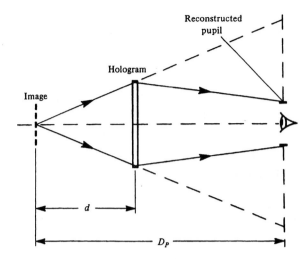

Fig. 2.5. Image reconstruction with an external reconstructed pupil [Hariharan, 1978].

object beam. As shown in fig. 2.5, the flux from the reconstructed image is confined within this external reconstructed pupil.

As a result, the solid angle over which the image can be viewed is reduced, but the luminance of the image is increased. A substantial improvement in image luminance can be obtained if the aperture of the imaging system is masked so that its shape and size match the range of angles over which the hologram is actually to be viewed.

References

Goodman, J. W. (1967). Film grain noise in wavefront reconstruction imaging. *Journal of the Optical Society of America*, **57**, 493–502.

Goodman, J. W. (1975) Statistical properties of laser speckle patterns. In *Laser Speckle & Related Phenomena*, Topics in Applied Physics, vol. 9, ed. J. C. Dainty, pp. 9–75. Berlin: Springer-Verlag.

Hariharan, P. (1976). Longitudinal distortion in images reconstructed by reflection holograms. *Optics Communications*, **17**, 52–4.

Hariharan, P. (1978). Hologram recording geometry: its influence on image luminance. *Optica Acta*, **25**, 527–30.

Hopkins, H. H. (1950). *Wave Theory of Aberrations*. Oxford: The Clarendon Press.

McKechnie, T. S. (1975). Speckle reduction. In *Laser Speckle & Related Phenomena*, Topics in Applied Physics, vol. 9, ed. J. C. Dainty, pp. 123–70. Berlin: Springer-Verlag.

Meier, R. W. (1965). Magnification and third-order aberrations in holography. *Journal of the Optical Society of America*, **55**, 987–92.

Rotz, F. B. & Friesem, A. A. (1966). Holograms with non-pseudoscopic real images. *Applied Physics Letters*, **8**, 146–8.

Problems

2.1. A hologram is recorded using a pulsed ruby laser ($\lambda = 694$ nm), and illuminated with a He–Ne laser ($\lambda = 633$ nm) to view the image. The reference beam in the recording system appears to diverge from a point at a distance of 1 m from the hologram. How far from the hologram should the beam from the He–Ne laser be brought to a focus to ensure that the image is reconstructed with unit lateral magnification?

From (2.7), it follows that the condition for the image to be reconstructed with unit lateral magnification is

$$\mu z_P = z_R, \tag{2.23}$$

where z_P and z_R are the distances from the hologram of the sources used for reconstruction and recording, respectively, and μ is the ratio of their wavelengths. Accordingly, the beam from the He–Ne laser should be brought to a focus at a distance from the hologram

$$z_P = 1 \times \frac{694}{633} \text{ m}$$
$$= 1.096 \text{ m}. \tag{2.24}$$

2.2. A hologram recorded with an Ar^+ laser ($\lambda_1 = 514$ nm) and an interbeam angle of 30° is to be illuminated with green light from a high-pressure mercury vapor lamp which has a mean wavelength λ_2 of 546 nm and a spectral bandwidth $\Delta\lambda_2$ of about 5 nm. The lamp can be regarded as a source with an effective diameter of 5 mm and is placed at a distance of 1 m from the hologram. What is the maximum depth of the image that can be reconstructed with an acceptable value of the image blur?

The acceptable value of the image blur for a display is determined by the resolution of the eye, which is about 0.5 mrad, or 0.5 mm at a viewing distance of 1 m.

The total image blur is the sum of the contributions from the finite size of the source and its spectral bandwidth and, from (2.17) and (2.18), can be written as

$$\Delta x_{I,\text{Total}} = z_O \left(\frac{\Delta x_P}{z_P} \right) + z_O \left(\frac{x_P}{z_P} \right) \left(\frac{\Delta\lambda_2}{\lambda_2} \right). \tag{2.25}$$

In the present case, $\Delta x_p = 5$ mm, $z_p = 1$ m and $(x_p/z_p) = \tan 30° = 1/\sqrt{3}$, so that we have

$$0.5 \times 10^{-3} = z_0 \left[\frac{5}{1000} + \frac{1}{\sqrt{3}} \frac{5}{546} \right],$$

$$= z_0 [5 \times 10^{-3} + 5.293 \times 10^{-3}], \tag{2.26}$$

and

$$z_0 = 48.6 \text{ mm.} \tag{2.27}$$

It follows that the distance of any point in the image from the hologram should not exceed this value.

Note that, in this case, the contributions of the size of the source and its spectral bandwidth to the image blur are almost equal.

2.3. The image reconstructed by a hologram illuminated with a He–Ne laser ($\lambda = 633$ nm) is to be photographed with a camera having a lens with a focal length of 100 mm. We need to stop the lens down to get the maximum depth of focus in the image. What is the highest *f*-number that can be used for the speckle size in the image to be less than 0.01 mm?

From (2.20), the minimum radius of the lens aperture is

$$\rho_{\text{min}} = \frac{0.61 \times 633 \times 10^{-9} \times 100 \times 10^{-3}}{0.01 \times 10^{-3}} \text{ m}$$

$$= 3.86 \text{ mm,} \tag{2.28}$$

which would correspond to an *f*-number of $100/(2 \times 3.86) = 12.95$.

3

Thin and thick holograms

So far, we have treated a hologram recorded on a photographic film as equivalent, to a first approximation, to a grating of negligible thickness with a spatially varying transmittance. However, if the thickness of the recording medium is larger than the average spacing of the fringes, volume effects cannot be neglected. It is even possible, as mentioned in Section 1.7, to produce holograms in which the interference pattern that is recorded consists of planes running almost parallel to the surface of the recording material; such holograms reconstruct an image in reflected light.

In addition, with modified processing techniques, or with other recording materials, it is possible to reproduce the variations in the intensity in the interference pattern produced by the object and reference beams as variations in the refractive index, or the thickness, of the hologram. Accordingly, holograms recorded in a medium whose thickness is much less than the spacing of the interference fringes (thin holograms) can be classified as amplitude holograms and phase holograms.

Similarly, holograms recorded in thick media (volume holograms) can be subdivided into transmission amplitude holograms, transmission phase holograms, reflection amplitude holograms and reflection phase holograms.

In the next few sections we review the characteristics of these six types of holograms. For simplicity, we consider only gratings produced by the interference of two plane wavefronts.

3.1 Thin gratings

3.1.1 Thin amplitude gratings

The amplitude transmittance of a thin amplitude grating can be written as

$$t(x) = t_0 + \Delta t \cos Kx, \tag{3.1}$$

where t_0 is the average amplitude transmittance, Δt is the amplitude of the spatial variations of $t(x)$, and $K = 2\pi/\Lambda$, where Λ is the average spacing of the fringes. The maximum amplitude in each of the two diffracted orders is a fourth of that in the incident wave, so that the maximum diffraction efficiency is

$$\eta_{max} = 0.0625. \qquad (3.2)$$

3.1.2 Thin phase gratings

If the phase shift produced by the recording medium is proportional to the intensity in the interference pattern, the resulting recording can be regarded as a thin phase grating whose complex amplitude transmittance can be written as

$$\mathbf{t}(x) = \exp(-i\phi_0)\,\exp[-i\Delta\phi\cos(Kx)], \qquad (3.3)$$

where ϕ_0 is a constant phase factor, and $\Delta\phi$ is the amplitude of the phase variations. If we neglect this constant phase factor, the right-hand side of (3.3) can be expanded to obtain the relation

$$\mathbf{t}(x) = \sum_{n=-\infty}^{\infty} i^n\,J_n(\Delta\phi)\,\exp(inKx), \qquad (3.4)$$

where J_n is the Bessel function of the first kind, of order n. The incident beam is diffracted into a number of orders, with the diffracted amplitude in the nth order proportional to the value of the Bessel function $J_n(\Delta\phi)$, but only the first diffracted order contributes to the primary image. The diffraction efficiency of the phase grating can therefore be written as

$$\eta = J_1^2(\Delta\phi). \qquad (3.5)$$

As shown in fig. 3.1, the amplitude diffracted into the first order increases initially with the phase modulation and then decreases; the maximum value of the diffraction efficiency is

$$\eta_{max} = 0.339. \qquad (3.6)$$

3.2 Volume gratings

With a thick recording medium, the hologram is made up of layers corresponding to a periodic variation of transmittance or refractive index. If the two interfering wavefronts are incident on the recording medium from the same side, these layers are approximately perpendicular to its surface, and the hologram produces an image by transmission. However, it is also possible to have the two

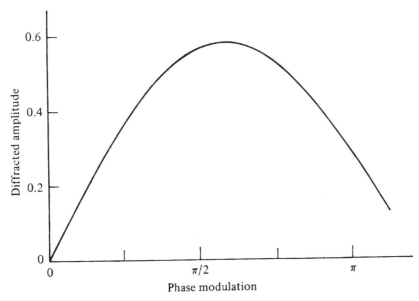

Fig. 3.1. Amplitude diffracted, as a function of the phase modulation, for a thin phase grating [Kogelnik, 1967].

interfering wavefronts incident on the recording medium from opposite sides (see Section 1.7), in which case the interference surfaces run approximately parallel to the surface of the recording medium. In this case, the reconstructed image is produced by the light reflected from the hologram.

We will use a coordinate system in which, as shown in fig. 3.2, the z axis is perpendicular to the surfaces of the recording medium, and the x axis is in the plane of incidence. For simplicity, we will assume that the interference surfaces are either perpendicular or parallel to the surfaces of the recording medium. The grating vector $\mathbf{K}$, which is perpendicular to the interference surfaces, is of length $K = 2\pi/\Lambda$, where Λ is the grating period, and makes an angle ψ ($\psi = 90°$ or $0°$ in this case) with the z axis.

In both cases, the diffracted amplitude is a maximum only when the angle of incidence is equal to the Bragg angle θ_B, satisfying the Bragg condition

$$\cos(\psi - \theta_B) = \frac{K}{2n_0 k_0}, \tag{3.7}$$

where n_0 is the average refractive index of the recording medium, and $k_0 = 2\pi/\lambda$. As a result, only two waves need be considered; they are the incoming reference wave R and the outgoing signal wave S. Since the other diffracted orders violate the Bragg condition strongly, they are severely attenuated and can be neglected. With volume reflection holograms, the angular and wavelength selectivity can

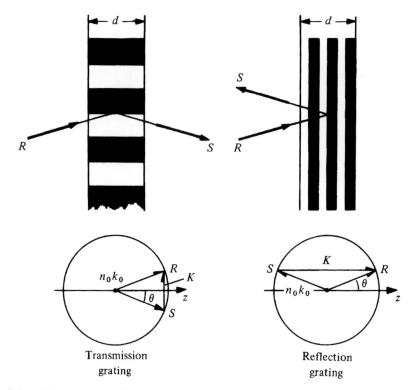

Fig. 3.2. Volume transmission and reflection gratings and their associated vector diagrams for Bragg incidence [Kogelnik, 1967].

be high enough to produce a sharp, monochromatic image when the hologram is illuminated with white light.

When analyzing the diffraction of light by volume gratings, it is necessary to take into account the fact that the amplitude of the diffracted wave increases progressively, while that of the incident wave decreases, as they propagate through the grating. This problem was solved by the development of a coupled wave theory [Kogelnik, 1969; Solymar & Cooke, 1981]. Some of the most important results for volume gratings are summarized below.

3.2.1 Volume transmission gratings

We consider, in the first instance, a lossless, volume transmission phase grating of thickness d, with the grating planes running normal to its surface. If we assume that the refractive index varies sinusoidally with an amplitude Δn about a mean value n_0, the diffraction efficiency of the grating at the Bragg angle θ_B is

$$\eta_B = \sin^2 \Phi, \tag{3.8}$$

where $\Phi = \pi \, \Delta n \, d / \lambda \cos \theta_B$ is known as the modulation parameter. The diffraction efficiency increases initially as the modulation parameter Φ is increased, until, when $\Phi = \pi/2$, $\eta_B = 1$. Beyond this point, the diffraction efficiency decreases.

For a deviation $\Delta \theta$ in the angle of incidence, or a deviation $\Delta \lambda$ in the wavelength of the incident beam, from the values required to satisfy the Bragg condition, the diffraction efficiency drops to

$$\eta = \frac{\sin^2 (\Phi^2 + \chi^2)^{1/2}}{(1 + \chi^2 / \Phi^2)}, \tag{3.9}$$

where

$$\chi = \Delta \theta \frac{Kd}{2}, \tag{3.10}$$

or, alternatively,

$$\chi = \frac{\Delta \lambda \, K^2 d}{8 \pi n_0 \cos \theta_B}. \tag{3.11}$$

Figure 3.3 shows the normalized diffraction efficiency (η / η_B) of a volume transmission phase grating, as a function of the parameter χ, for three values of the modulation parameter Φ.

The other case we shall consider is that of a volume transmission grating in which the refractive index does not vary, but the absorption constant varies with an amplitude $\Delta \alpha$ about its mean value α. In this case, the diffraction efficiency, for incidence at the Bragg angle, is given by the expression

$$\eta = \exp \left(\frac{-2 \alpha d}{\cos \theta_B} \right) \sinh^2 \left(\frac{\Delta \alpha \, d}{2 \cos \theta_B} \right). \tag{3.12}$$

The maximum diffraction efficiency is obtained when

$$\Delta \alpha = \alpha = \frac{\ln 3}{d \cos \theta_B} \tag{3.13}$$

and has a value $\eta_{max} = 0.037$.

3.2.2 Volume reflection gratings

The diffraction efficiency of a volume reflection phase grating at the Bragg angle is given by the relation

$$\eta_B = \tanh^2 \Phi_r, \tag{3.14}$$

where $\Phi_r = \pi \, \Delta n \, d / \lambda \cos \theta_B$ and Δn is the amplitude of the variation in the refractive index. As the value of Φ_r increases, the diffraction efficiency increases steadily, as shown in fig. 3.4, to a limiting value of 1.00.

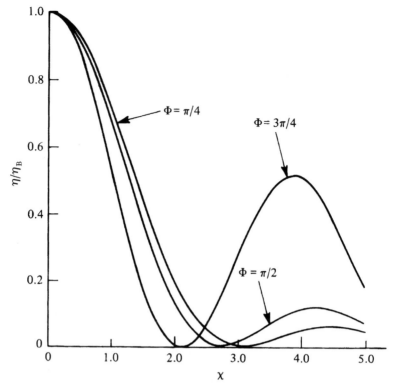

Fig. 3.3.   Normalized diffraction efficiency of a volume transmission phase grating as a function of the parameter χ, which is a measure of the deviation from the Bragg condition, for three values of the modulation parameter Φ [Kogelnik, 1969].

For a deviation from the Bragg condition, the normalized diffraction efficiency decreases, as shown in fig. 3.5, as a function of the parameter χ_r specified by the relations

$$\chi_r = \Delta\theta\left(\frac{2\pi n_0 d}{\lambda}\right)\sin\theta_B \qquad (3.15)$$

$$= \left(\frac{\Delta\lambda}{\lambda}\right)\left(\frac{2\pi n_0 d}{\lambda}\right)\cos\theta_B. \qquad (3.16)$$

The diffraction efficiency drops to zero for a value of $\chi_r \approx 3.5$.

3.3 Imaging properties

With volume holograms, which exhibit a high degree of angular and wavelength selectivity, the amplitude of the diffracted wavefront is affected by any changes of wavelength or geometry between recording and reconstruction.

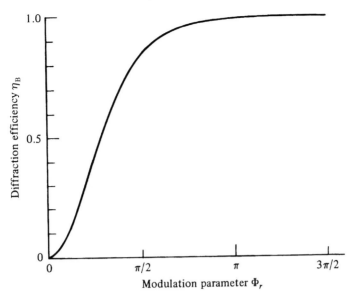

Fig. 3.4. Diffraction efficiency of a volume reflection phase grating as a function of the modulation parameter Φ_r.

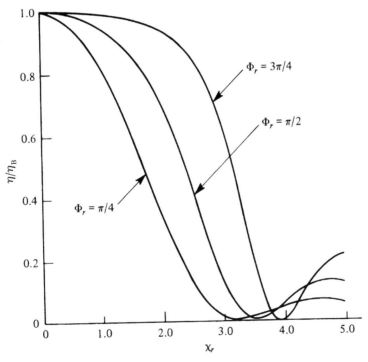

Fig. 3.5. Normalized diffraction efficiency (η/η_B) of a volume reflection phase grating as a function of the parameter χ_r, which is a measure of the deviation from the Bragg condition, for different values of the modulation parameter Φ_r [Kogelnik, 1969].

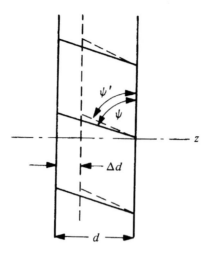

Fig. 3.6. Changes in the orientation and spacing of the fringe planes in a volume transmission hologram due to a change in the thickness of the photographic emulsion.

Such effects are particularly noticeable with holograms recorded on photographic materials, where processing usually results in a change in the thickness of the photographic emulsion.

In the case of a reflection hologram, such a change in thickness results in a change in the spacing of the interference surfaces and, consequently, a change in the color of the image formed when the hologram is illuminated with white light.

With transmission holograms, a change in thickness can result, as shown in fig. 3.6, in a rotation of the fringe planes as well as a change in their spacing [Vilkomerson & Bostwick, 1967; Hariharan, 1999]. For gratings recorded with plane wavefronts, it is possible to compensate for these changes by changing either the angle of illumination or the wavelength. However, complete compensation is not possible with a hologram of a point at a finite distance or a hologram of an extended object.

3.4 Thick and thin gratings

The distinction between thin gratings and volume gratings is commonly made [Klein & Cook, 1967] on the basis of a parameter Q defined by the relation

$$Q = \frac{2\pi\lambda d}{n_0 \Lambda^2}.$$

(3.17)

Small values of Q $(Q<1)$ correspond to thin gratings, while large values of Q $(Q>1)$ correspond to volume gratings. However, more detailed studies have

Table 3.1. *Theoretical maximum diffraction efficiencies for transmission phase holograms*

Hologram type	Thin		Volume	
Object beam	Collimated	Diffuse	Collimated	Diffuse
η_{max}	0.33	0.22	1.00	0.64

shown that the transition between the two regimes is not completely defined by (3.17) and that, as the modulation amplitude increases, an intermediate regime appears and widens [Moharam, Gaylord & Magnusson, 1980*a*,*b*].

3.5 Diffusely reflecting objects

The values of diffraction efficiency obtained with a hologram of a diffusely reflecting object are always much lower than those for a grating, because it is not possible to maintain optimum modulation over the entire area, due to the local variations in the amplitude of the object wave. The maximum diffraction efficiencies of transmission phase holograms recorded with a diffuse object beam have been calculated [Upatnieks & Leonard, 1970] on the assumption that the amplitude of the object wave has statistics similar to those of a speckle pattern and are presented in Table 3.1.

3.6 Multiply exposed holograms

With a thick recording medium, it is possible to record two or more holograms in the same medium and read them out separately. In order to do this, their Bragg angles should be sufficiently far apart that the maximum of the angular selectivity curve for one hologram coincides with the first minimum for the other. However, with N amplitude transmission holograms, the diffraction efficiency of each hologram drops to $1/N^2$ of that for a single hologram, since the available dynamic range is divided equally between the N holograms.

On the other hand, with volume phase holograms, if the Bragg angles are far enough apart for coupling between the gratings to be negligible, each hologram diffracts independently of the others [Case, 1975], and there is no loss in diffraction efficiency. However, if the recording medium is nearing saturation, the consequent reduction in modulation can result in a decrease in the diffraction efficiencies of the individual holograms.

References

Case, S. K. (1975). Coupled wave theory for multiply exposed thick holograms. *Journal of the Optical Society of America*, **65**, 724–9.

Hariharan, P. (1999). Real-time holographic interferometry: effects of emulsion shrinkage. *Optics & Lasers in Engineering*, **31**, 339–44.

Klein, W. R. & Cook, B. D. (1967). Unified approach to ultrasonic light diffraction. *IEEE Transactions on Sonics & Ultrasonics*, **SU-14**, 123–34.

Kogelnik, H. (1967). Reconstructing response and efficiency of hologram gratings. In *Proceedings of the Symposium on Modern Optics*, pp. 605–17. Brooklyn Polytechnic Press.

Kogelnik, H. (1969). Coupled wave theory for thick hologram gratings. *Bell System Technical Journal*, **48**, 2909–47.

Moharam, M., Gaylord, T. K. & Magnusson, R. (1980a). Criteria for Bragg regime diffraction by phase gratings. *Optics Communications*, **32**, 14–8.

Moharam, M., Gaylord, T. K. & Magnusson, R. (1980b). Criteria for Raman-Nath regime diffraction by phase gratings. *Optics Communications*, **32**, 19–23.

Solymar, L. & Cooke, D. J. (1981). *Volume Holography & Volume Gratings*. New York: Academic Press.

Upatnieks, J. & Leonard, C. (1970). Efficiency and image contrast of dielectric holograms. *Journal of the Optical Society of America*, **60**, 297–305.

Vilkomerson, D. H. R. & Bostwick, D. (1967). Some effects of emulsion shrinkage on a hologram's image space. *Applied Optics*, **6**, 1270–2.

Problems

3.1. A volume transmission phase grating with a spatial frequency of 1579 lines/mm (see Problem 1.1), recorded in a gelatin layer with a thickness of 15 μm, has a diffraction efficiency of 100 percent at the Bragg angle. What would be the angular selectivity of the grating at a wavelength of 633 nm?

The spacing of the grating fringes is $\Lambda = (1/1579)$ mm $= 0.633$ μm, so that the value of the grating vector $K = 2\pi/\Lambda = 9.926 \times 10^6$ m^{-1}. Since the diffraction efficiency of the grating at the Bragg angle is 100 percent, the modulation parameter $\Phi = \pi/2$. From the curve in fig. 3.3 corresponding to this value of Φ, the diffraction efficiency of the grating drops to zero when the dephasing parameter $\chi = 2.7$. Accordingly, from (3.10), the deviation from the Bragg angle at which the diffraction efficiency drops to zero is

$$\Delta\theta = 2\chi/Kd$$

$$= \frac{2 \times 2.7}{9.926 \times 10^6 \times 15 \times 10^{-6}} \text{ radian}$$

$$= 3.63 \times 10^{-2} \text{ radian}$$

$$= 2.08°. \tag{3.18}$$

3.2. A volume reflection phase grating (see Problem 1.2) is recorded in a photographic emulsion layer with a thickness of 15 μm. What would be the wavelength selectivity of this grating at a mean wavelength of 633 nm?

Within the emulsion layer, the effective wavelength is 395.6 nm and the Bragg angle is $90° - 76.85° = 13.15°$. From the curves in fig. 3.5, the diffraction efficiency drops to zero when $\chi_r = 3.5$. From (3.16), this would correspond to a change in wavelength

$$\Delta\lambda = \frac{3.5(395.6 \times 10^{-9})^2}{2\pi \times 1.6 \times 15 \times 10^{-6} \cos 13.15°} \text{ m}$$
$$= 3.73 \text{ nm}, \qquad (3.19)$$

in the emulsion, or a change in wavelength of 5.97 nm in air.

4

Light sources

In order to maximize the visibility of the interference fringes formed by the object and reference beams, while recording a hologram, it is essential to use coherent illumination (see Appendix A). In addition to being spatially coherent, the coherence length of the light must be much greater than the maximum value of the optical path difference between the object and the reference beams in the recording system. Lasers are therefore employed almost universally as light sources for recording holograms.

4.1 Lasers

The characteristics of some of the lasers used for holography are listed in Table 4.1.

For a simple holographic system, the helium–neon (He–Ne) laser is the usual choice. It is inexpensive and operates on a single spectral line at 633 nm which is well matched to the peak sensitivity of many photographic emulsions. In addition, it does not require water cooling and has a long life.

Table 4.1. *Lasers used for holography*

Laser	Output	Wavelength (nm)	Power
Ar^+	cw	514, 488	1 W
He–Cd	cw	442	25 mW
He–Ne	cw	633	2–50 mW
Kr^+	cw	647	500 mW
Diode	cw	670–650	5 mW
Diode–YAG	cw	532	100 mW
Dye	cw	tunable	200 mW
Ruby	pulsed	694	1–10 J

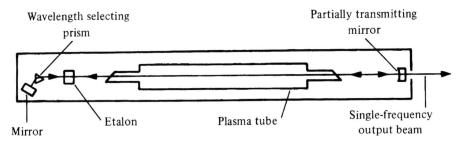

Fig. 4.1. Optical system of an argon-ion laser.

Argon-ion (Ar^+) lasers are more expensive and complex, but give much higher outputs in the green (514 nm) and blue (488 nm) regions of the spectrum. The Ar^+ laser normally has a multiline output but can be made to operate at a single wavelength by replacing the reflecting end mirror by a prism and mirror assembly, as shown in fig. 4.1. The krypton-ion (Kr^+) laser is useful where high output power is required at the red end of the spectrum (647 nm).

The helium–cadmium (He–Cd) laser provides a stable output at a relatively short wavelength (442 nm) and is useful with recording materials such as photoresists.

Diode lasers can be used as a source of red light, and are now available with output wavelengths that are fairly well matched to the sensitivity of commercial photographic materials.

Diode lasers can also be used to pump a Nd:YAG laser with a frequency doubler. Such a system provides a compact cw source of green light ($\lambda = 532$ nm) with output powers up to 100 mW.

Dye lasers are relatively expensive and complex, but have the advantage that they can be operated over a wide range of wavelengths by switching dyes; in addition, for applications such as contouring (see Section 14.3), the output can be tuned over a range of wavelengths (50–80 nm) by incorporating a wavelength selector, such as a diffraction grating or a birefringent filter, in the laser cavity.

While pulsed Nd:YAG lasers (with a frequency doubler) have been used for holography, the most commonly used type of pulsed laser is the ruby laser, mainly because of the large output energy available (up to 10 J per pulse) and the wavelength of the light emitted (694 nm), which is fairly well matched to the peak sensitivity of available photographic materials for holography.

The active medium in a ruby laser is a rod, typically 5–10 mm in diameter and 75–100 mm in length, made of synthetic sapphire (Al_2O_3) doped with 0.05 percent of Cr_2O_3. When this rod is mounted in an optical resonator, as shown

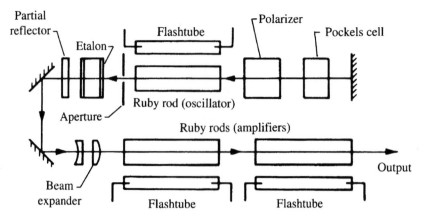

Fig. 4.2.   Optical system of a *Q*-switched ruby laser.

in fig. 4.2, and pumped with a xenon flash lamp, it emits a series of pulses [Lengyel, 1971]. Single pulses are obtained by using a *Q*-switch (a Pockels cell) in the cavity.

Since the output from a ruby oscillator is limited, high-power ruby lasers use one or more additional ruby rods as amplifiers. With two stages of amplification, an output of 10 J can be obtained.

4.2 Coherence requirements

As mentioned earlier, to obtain a satisfactory hologram, the light used must be spatially coherent and its coherence length (see Appendix A) must be much greater than the maximum optical path difference between the object and reference beams in the recording system.

Spatial coherence is automatically ensured if the laser oscillates in the lowest order transverse mode (the TEM_{00} mode); this mode also gives the most uniform illumination.

Operation of Ar^+ and Kr^+ lasers on a single spectral line can be obtained, where necessary, by means of a wavelength selector prism. However, this may not ensure adequate temporal coherence since many lasers, even when operating on a single spectral line, will oscillate, as shown in the upper part of fig. 4.3, in a number of longitudinal modes lying within the profile of this spectral line, at which the gain of the laser medium is adequate to overcome the cavity losses. These modes correspond to the resonant frequencies of the laser cavity and are separated by a frequency interval

$$\Delta v = c/2L, \tag{4.1}$$

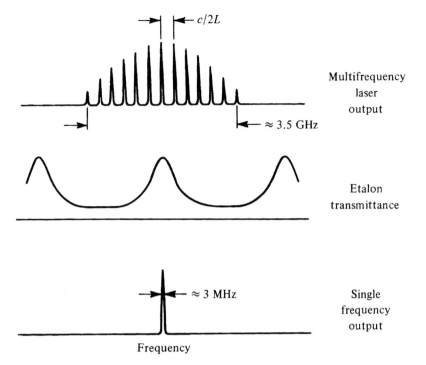

Fig. 4.3. Laser modes, without and with an intracavity etalon.

where c is the speed of light, and L is the length of the laser cavity. If we assume that the output power is divided equally between N longitudinal modes, the effective coherence length of the output is

$$\Delta l = 2L \big/ N. \qquad (4.2)$$

Equation (4.2) shows that the existence of more than one longitudinal mode in the output reduces the coherence length severely. Even if the mean optical paths of the object and reference beams are equalized carefully, severe restrictions are placed on the maximum depth of the object.

Depending on their power, commercial He–Ne lasers oscillate, typically, in two to five longitudinal modes, and the coherence length of the output is limited to a few centimeters. With high-power Ar$^+$ and Kr$^+$ lasers, it is possible to obtain operation in a single longitudinal mode, and coherence lengths in excess of a meter, by using an intracavity etalon. This etalon is tuned to obtain maximum power output, and stabilize the output wavelength, by mounting it in a temperature-controlled oven.

Diode lasers can be made to operate in a single longitudinal mode by increasing the drive current to a figure just below their maximum rating [Henry, 1991].

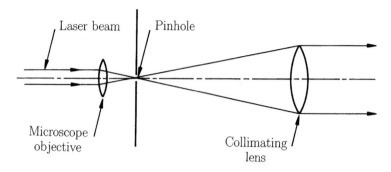

Fig. 4.4.   Optical system used to expand and spatially filter a laser beam.

4.3 Laser beam expansion

Since the beam from most lasers has a diameter of only 1–2 mm and a very low divergence, it is usually necessary to use low-power microscope objectives to expand the laser beam to illuminate the object as well as the hologram. A problem is that, due to the high coherence of laser light, the expanded beam exhibits random diffraction patterns (spatial noise) produced by defects and dust on the optical surfaces in the path of the beam. These diffraction patterns can be eliminated, and a clean beam obtained, by placing a pinhole at the focus of the microscope objective, as shown in fig. 4.4.

If the laser is oscillating in the TEM_{00} mode, the beam has a Gaussian intensity profile given by the relation

$$I(r) = I(0) \, \exp\!\left(\frac{-2r^2}{w^2}\right), \tag{4.3}$$

where r is the radial distance from the center of the beam and w is the distance at which the intensity drops to $(1/e^2)$ of that at the center of the beam. Since the aperture of the microscope objective is usually greater than $2w$, the diameter of the focal spot is

$$d = \frac{2\lambda f}{\pi w}, \tag{4.4}$$

where f is the focal length of the microscope objective. If the diameter of the pinhole is less than d, randomly diffracted light is blocked, and the transmitted beam has a smooth profile.

An exceptional case, where it is possible to dispense with beam expansion, is with a diode laser. Because of the extremely small emitting area, a diode laser produces a clean, divergent beam similar to that obtained from a He–Ne laser after expansion and spatial filtration. Diode lasers can therefore be used to record some types of holograms (see Section 7.2) without any external optics.

4.4 Beam splitters

If we define the beam ratio $R = (r/o)^2$ as the ratio of the irradiances of the reference and object beams, we would expect the visibility of the interference fringes forming the hologram to be a maximum when $R = 1$. However, the wave scattered by a diffusely reflecting object exhibits very strong local variations in amplitude (see Section 2.5). As a result it is usually necessary to work with a value of $R \gg 1$ (typically $R \approx 3$), to avoid nonlinear effects.

A convenient way to vary the ratio of the powers in the two beams and optimize the visibility of the hologram fringes is to use a beam splitter consisting of a glass disc, which can be rotated, coated with a thin aluminum film whose reflectivity is a linear function of the azimuth. Such a variable-ratio beam splitter must be used in the unexpanded laser beam to minimize the variation in reflectivity across the beam.

4.5 Beam polarization

Most gas lasers have Brewster-angle windows on the plasma tube so that the output is linearly polarized. The visibility of the interference fringes forming the hologram is then a maximum when the electric vectors of the object and reference beams are parallel (see Appendix A). This condition is always satisfied if the two beams are linearly polarized with their electric vectors normal to the plane containing the beams. If, on the other hand, they are polarized with their electric vectors in the plane containing the beams, the visibility of the hologram fringes can drop to zero when the beams intersect at right angles.

It should be noted that in the case of an object with a rough surface, a substantial fraction of the reflected light may be depolarized. The resulting decrease in the visibility of the interference fringes can be minimized, where necessary, by using a sheet polarizer in front of the hologram to eliminate the cross-polarized component.

4.6 Pulsed laser holography

Very short light pulses (<20 nanoseconds) can be obtained with a pulsed laser, if a Pockels cell is used as a Q-switch in the laser cavity. As a result, problems of vibration and air currents are largely eliminated. Because their output

Fig. 4.5.   Professor Gabor with his holographic portrait: this hologram was produced by R. Rinehart at the McDonnell Douglas Electronics company in 1971, using a pulsed ruby laser.

wavelength is well matched to the peak sensitivity of available photographic materials, and their output energy is fairly large, pulsed ruby lasers are used widely to record holograms in an industrial environment [Koechner, 1979].

A Q-switched ruby laser can also be used to record holograms of living human subjects [Siebert, 1968] as shown in fig. 4.5. However, to avoid eye damage (see Section 4.7), instead of illuminating the subject directly, the expanded laser beam should be allowed to fall on a large diffuser which constitutes an extended source illuminating the subject.

A studio setup for making holographic portraits has been described by Bjelkhagen [1992].

4.7 Laser safety

Since the beam from a laser is focused by the lens of the eye to a very small spot on the retina, direct exposure to low power lasers can cause eye damage. With pulsed lasers, even stray reflections can be dangerous. It is essential to take all due precautions to avoid accidental exposure and, where required, to use appropriate eye protection [Sliney & Wolbarsht, 1980].

References

Bjelkhagen, H. I. (1992). Holographic portraits made by pulse lasers. *Leonardo*, **25**, 443–8.

Henry, C. H. (1991). Theory of the linewidth of semiconductor lasers. In *Semiconductor Diode Lasers*, Vol. 1, pp. 36–41. New York: IEEE.

Koechner, W. (1979). Solid state lasers. In *Handbook of Optical Holography*, ed. H. J. Caulfield, pp. 257–67. New York: Academic Press.

Lengyel, B. A. (1971). *Lasers*. New York: Wiley-Interscience.

Siebert, L. D. (1968). Large scene front-lighted hologram of a human subject. *Proceedings of the IEEE*, **56**, 1242–3.

Sliney, D. & Wolbarsht, M. (1980). *Safety with Lasers and Other Optical Sources: a Comprehensive Handbook*. New York: Plenum Press.

Problems

4.1. A He–Ne laser with a 200 mm long resonant cavity oscillates in two longitudinal modes. What is the coherence length of the radiation?

From (4.2) the coherence length of the radiation is

$$\Delta l = 2L/N$$
$$= (2 \times 0.2/2) \text{ m}$$
$$= 0.2 \text{ m.} \tag{4.5}$$

4.2. In the arrangement shown in fig. 4.4, the central part of the beam from a He–Ne laser is isolated by an aperture with a diameter of 2.0 mm and brought to a focus by a microscope objective with a focal length of 32 mm. What would be a suitable size for the pinhole?

The diameter of the focal spot is equal to the diameter of the Airy disc which, in this case, is

$$d = \frac{2.44 \times 0.633 \times 10^{-6} \times 32 \times 10^{-3}}{2 \times 10^{-3}} \text{ m}$$
$$= 24.4 \text{ μm.} \tag{4.6}$$

A pinhole with a diameter of 20 μm would ensure a clean beam with only a marginal loss of light.

5

The recording medium

5.1 Amplitude and phase holograms

The complex amplitude transmittance of the medium used to record a hologram can be written as

$$\mathbf{t} = \exp(-\alpha d)\,\exp[-i(2\pi n d/\lambda)],$$
$$= |\mathbf{t}|\,\exp(-i\phi),\qquad\qquad(5.1)$$

where α is the absorption constant of the medium, d is its thickness and n is its refractive index. A change in α with the exposure produces an amplitude hologram, while a change in n or d will produce a phase hologram.

The response of a recording material used for amplitude holograms can be characterized on a macroscopic scale, by a curve of the amplitude transmittance as a function of the exposure (a $|\mathbf{t}| - E$ curve), as shown in fig. 5.1.

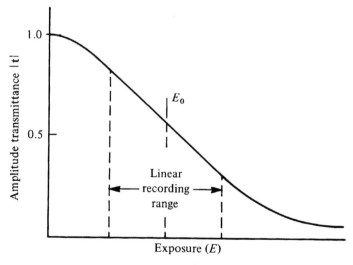

Fig. 5.1.   Typical amplitude transmittance vs. exposure curve for a recording material.

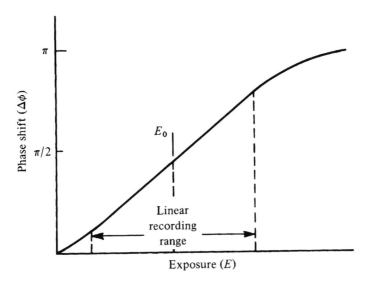

Fig. 5.2. Typical phase shift vs. exposure curve for a recording material.

Similarly, the response of a recording material used for phase holograms can be described by a curve of the effective phase shift as a function of the exposure (a $\Delta\phi - E$ curve), as shown in fig. 5.2.

5.2 The modulation transfer function

Curves such as those shown in figs. 5.1 and 5.2 may not describe the response of the recording medium on a microscopic scale because the actual intensity distribution to which the recording medium is exposed always differs from that incident on it, due to scattering and absorption in the medium. As a result, the actual modulation of the intensity within the recording medium is always less than that in the original interference pattern. In addition, the response of the medium at different spatial frequencies may be affected by the processing technique that is used.

For a given modulation of the input intensity, the ratio of the response of the recording medium at any spatial frequency s, relative to that at low spatial frequencies $(s\rightarrow 0)$, is specified by a parameter $M(s)$, termed the modulation transfer function.

5.3 Effects of nonlinearity

For simplicity, we have assumed so far that the amplitude transmittance of the hologram is a linear function of the intensity, as described by (1.2). However, in practice, this assumption is not always valid. The amplitude transmittance of the processed recording material can then be represented by a polynomial [Bryngdahl & Lohmann, 1968]

$$\begin{aligned}
\mathbf{t} &= \mathbf{t_0} + \beta_1 TI + \beta_2 T^2 I^2 + \cdots \\
&= \mathbf{t_0} + \beta_1 T[rr^* + oo^* + r^*o + ro^*] \\
&\quad + \beta_2 T^2[rr^* + oo^* + r^*o + ro^*]^2 \\
&\quad + \cdots.
\end{aligned} \tag{5.2}$$

If the hologram is illuminated once again with a plane wave of unit amplitude, the complex amplitude of the wave transmitted by the hologram can be written in the form

$$\begin{aligned}
u &= \text{linear terms} \\
&\quad + \beta_2 T^2[(oo^*)^2 + o^2 + o^{*2} + 2o^2o^* + 2oo^{*2}] \\
&\quad + \cdots.
\end{aligned} \tag{5.3}$$

As can be seen, nonlinearity leads to the production of additional spurious terms. An examination of (5.3) shows that the term involving $(oo^*)^2$ results in a doubling of the width of the halo surrounding the directly transmitted beam, while the terms involving o^2 and o^{*2} correspond to higher-order diffracted images, and the terms involving $2o^2o^*$ and $2oo^{*2}$ are intermodulation terms, producing false images.

The effects of nonlinearity are particularly noticeable with phase holograms. Even if we assume that the phase shift produced by the recording medium is proportional to the exposure, the complex amplitude transmittance is given by the expression

$$\begin{aligned}
\mathbf{t} &= \exp(-i\phi) \\
&= 1 - i\phi - (1/2)\phi^2 + (1/6)i\phi^3 + \cdots.
\end{aligned} \tag{5.4}$$

If the phase modulation is increased, to obtain better diffraction efficiency, the effects of the higher-order terms cannot be neglected.

However, with volume holograms, the effects of nonlinearity are reduced significantly by the angular selectivity of the hologram. If the angle between the beams in the recording setup is large enough that the diffracted beams corresponding to different orders do not overlap, a simple analysis [Hariharan, 1979] shows that the signal-to-noise ratio should improve by a factor approximately equal to $(\psi/\Delta\theta)$, where $2\Delta\theta$ is the width of the passband of the angular selectivity function, and ψ is the angle subtended by the object at the hologram.

References

Bryngdahl, O. & Lohmann, A. (1968). Nonlinear effects in holography. *Journal of the Optical Society of America*, **58**, 1325–34.

Hariharan, P. (1979). Intermodulation noise in amplitude holograms: the effect of hologram thickness. *Optica Acta*, **26**, 211–5.

Problems

5.1. A hologram is recorded in an optical system in which the object and reference beams make angles of $+30°$ and $-30°$, respectively, with the normal to the photographic film. After exposure, the film is processed to produce a volume phase hologram. If the object subtends an angle of 30° at the photographic film, and the thickness of the photographic emulsion layer is 15 μm, what is the improvement in the signal-to-noise ratio due to the angular selectivity of the hologram?

Since the modulation parameter for such a hologram would be significantly lower than the optimum value of $\pi/2$ (say, $\pi/4$), it follows from fig. 3.3 that the diffraction efficiency drops to zero when the dephasing parameter $\chi \approx 3$. From Problems 1.1 and 3.1, this would correspond to a deviation from the Bragg angle $\Delta\theta = \pm 2.31°$. The signal-to-noise ratio should improve by a factor of $(30/4.62) \approx 6.5$.

6

Recording materials

Several recording materials have been used for holography [Smith, 1977]. Table 6.1 lists the principal characteristics of those that have been found most useful.

6.1 Photographic materials

High-resolution photographic plates and films were the first materials used to record holograms. They are still used widely because of their relatively high sensitivity when compared to other hologram recording materials [Bjelkhagen, 1993]. In addition, they can be dye sensitized so that their spectral sensitivity matches the most commonly used laser wavelengths.

Conventional processing produces an amplitude hologram. It also results in a reduction in the thickness of the emulsion layer of about 15 percent, due to the removal of the unexposed silver halide grains in the fixing bath. This reduction in thickness can cause a rotation of the fringe planes as well as a reduction in their spacing (see Section 3.3).

With volume reflection holograms (see Section 3.2.2), the reduction in the

Table 6.1. *Recording materials for holography*

Material	Exposure $(\mathrm{J/m^2})$	Resolution $(\mathrm{mm^{-1}})$	Processing	Type	η_{max}
Photographic	≈ 1.5	≈ 5000	Normal	Amplitude	0.06
			Bleach	Phase	0.60
DCG	10^2	10000	Wet	Phase	0.90
Photoresists	10^2	3000	Wet	Phase	0.30
Photopolymers	$10\text{--}10^4$	5000	Dry	Phase	0.90
PTP	10^{-1}	500–1200	Dry	Phase	0.30
BSO	10	10000	None	Phase	0.20

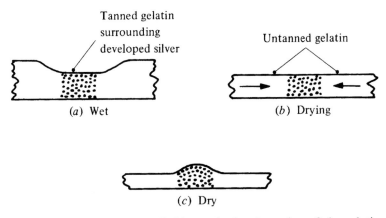

Fig. 6.1. Formation of a relief image by local tanning of the gelatin.

spacing of the fringe planes is immediately apparent as a shift in the color of the image, which is reconstructed at a shorter wavelength than that used to record it.

With volume transmission holograms, emulsion shrinkage can result in a reduction in diffraction efficiency when the hologram is replaced in the original recording setup, since the reference beam is no longer incident on it at the Bragg angle (see Section 3.2.1). This loss in diffraction efficiency can be minimized by having the object and reference beams incident at equal, but opposite, angles on the hologram, so that the fringe planes are normal to the surface of the photographic emulsion.

Amplitude holograms recorded in photographic emulsions can also exhibit phase modulation due to a surface relief image arising from local tanning (hardening) of the gelatin by the oxidation products of the developer [Smith, 1968]. After developing and fixing, as shown in fig. 6.1, the unexposed areas of the wet emulsion layer, which are not tanned, are swollen to more than five times their normal thickness and are very soft. The tanned gelatin in the exposed areas absorbs less water and, therefore, dries more quickly. Shrinkage during drying pulls some of the soft gelatin from adjacent unexposed areas into the exposed areas, so that, when the emulsion has dried, the exposed areas stand out in relief.

Since this relief image is confined to low spatial frequencies (≤ 200 mm^{-1}), it normally contributes very little to the reconstructed image, but can give rise to intermodulation terms.

Because of the low diffraction efficiency of amplitude holograms, holograms produced for displays using photographic materials are usually processed to obtain volume phase holograms (see Section 3.2) which have much higher

Developed

Ag   AgH   Ag

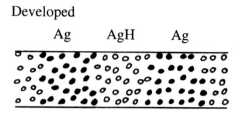

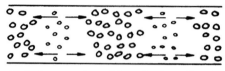

Rehalogenating bleach (no fixer)

Fig. 6.2. Production of photographic volume phase holograms by a rehalogenating bleach.

diffraction efficiencies. The most common procedure is to use a rehalogenating bleach bath, without fixing, to convert the developed silver back into a transparent silver halide with a high refractive index [Hariharan, Ramanathan & Kaushik, 1971; Phillips *et al.*, 1980]. The regenerated silver halide goes into solution and is redeposited on the adjacent unexposed silver halide grains. The result is that, as shown in fig. 6.2, a volume phase hologram is produced by material transfer from the exposed areas to the adjacent unexposed areas [Hariharan, 1990].

Material transfer through diffusion is effective only over a very limited distance. As a result, the diffraction efficiency of the hologram drops off rapidly, as shown in fig. 6.3, for fringe spacings greater than a critical value corresponding to the diffusion length of the silver ion in the bleach bath [Hariharan & Chidley, 1988].

Since, with a rehalogenating bleach, little or no silver or silver halide is removed from the emulsion layer, the reduction in the thickness of the emulsion due to processing is minimal. In the case of a volume reflection hologram, the change in the color of the reconstructed image is quite small, even with a nontanning developer, while the use of a tanning developer may result in an increase in thickness, so that the image is reconstructed at a longer wavelength [Hariharan & Chidley, 1989].

Until a few years ago, the most commonly used photographic plates and films were Kodak 649F and Agfa 8E75HD and 8E56HD. After the

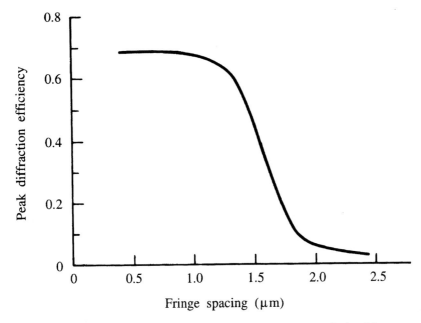

Fig. 6.3. Peak diffraction efficiency, plotted as a function of the fringe spacing, for transmission gratings produced in a photographic emulsion processed with a rehalogenating bleach [Hariharan & Chidley, 1988].

manufacture of these materials was discontinued, their place has been taken by the BB emulsions produced by Holographic Recording Technologies and the Slavich emulsions marketed by Geola.

The BB series of plates includes three emulsions (BB-640, BB-520 and BB-450) sensitized, respectively, for red, green and blue laser light. They have a grain size of 20–25 nm and require an exposure of approximately $1.5 \, J/m^2$ to obtain a density of 2.5 when developed in a metol-ascorbate developer. An emulsion with panchromatic sensitization (BB-PAN) is also available for making full-color reflection holograms (see Section 8.1), as well as one (BB-700) for use with pulsed lasers.

The Slavich series of films and plates features two emulsions (PFG-01 and VRP-M, sensitized for red and green laser light, respectively), which are similar to Agfa 8E75HD and 8E56HD, as well as two emulsions (PFG-03M, sensitized for the red, and PFG-03C, with panchromatic sensitization) with ultra-fine grains. The latter two emulsions require exposures of 15–20 J/m^2 but can resolve more than 5000 lines/mm.

Information on the BB emulsions is available at http://www.hrt-gmbh.de, and on the Slavich plates and films at http://www.geola.com.

6.2 Dichromated gelatin

A volume phase hologram can be recorded in a gelatin layer containing a small amount of a dichromate, such as $(NH_4)_2Cr_2O_7$, by making use of the fact that dichromated gelatin (DCG) becomes locally hardened on exposure to light, due to the formation of cross-links between the carboxylate groups on neighboring gelatin chains. This effect is used to obtain a local modulation of the refractive index.

After exposure in the holographic system using blue light from an Ar^+ laser ($\lambda = 488$ nm), the gelatin layer is washed in water at 20–30 °C for 10 min, so that it absorbs water and swells. The swollen gelatin layer is then immersed in two successive baths of isopropanol, to extract the water, and dried thoroughly. With care in processing, volume phase holograms with high diffraction efficiency and low scattering can be produced [Chang & Leonard, 1979; Jeong, Song & Lee, 1991].

A method commonly used to prepare plates coated with DCG is to dissolve and remove the silver halide from the emulsion layer in a photographic plate by soaking it in a nonhardening fixing bath. The plates are sensitized by soaking them for about 5 min at 20 °C in an aqueous solution of $(NH_4)_2Cr_2O_7$ to which a small amount of a wetting agent has been added. They are then allowed to drain and dried at 25–30 °C in darkness.

Geola also make plates (PFG-04) coated with dichromated gelatin which can be used directly to record reflection holograms. These plates require an exposure of $1000 \text{ J}/\text{m}^2$ at 488 nm and can yield gratings with a diffraction efficiency of 0.8.

6.3 Silver-halide sensitized gelatin

This technique makes it possible to combine the high sensitivity of photographic materials with the high diffraction efficiency, low scattering and high light-stability of DCG.

In this technique [Pennington, Harper & Laming, 1971], the exposed photographic emulsion is developed in a metol-hydroquinone developer and then bleached in a bath containing $(NH_4)_2Cr_2O_7$. During the bleaching process, the developed silver is oxidized to Ag^+, while the Cr^{6+} ions in the bleach are reduced to Cr^{3+} ions. The oxidation products of the developer, as well as the Cr^{3+} ions formed by reduction of the bleach, form cross-links between the gelatin chains in the vicinity of the oxidized silver grains, causing local hardening of the gelatin [Hariharan, 1986]. The emulsion is then fixed to remove the unexposed silver halide, washed, dehydrated with isopropanol and dried exactly as for a dichromated gelatin hologram. With this technique it is possible to obtain diffraction efficiencies up to 80 percent with transmission gratings and 55 percent with reflection gratings [Angell, 1987; Fimia, Pascual & Belendez, 1992].

Optimized processing techniques for SHSG holograms using BB-640 plates have been described by Belendez *et al.* [1998].

6.4 Photoresists

In positive photoresists, such as Shipley AZ-1350, the areas exposed to light become soluble and are washed away during development to produce a relief image [Bartolini, 1977].

The photoresist is coated on a glass substrate by spinning to form a layer 1–2 μm thick. This layer is then baked at 75 °C for 15 min to ensure complete removal of the solvent. Holograms are recorded with a He–Cd laser at a wavelength of 442 nm. The exposed plate is processed in AZ-303 developer diluted with four parts of distilled water.

Holograms recorded on a photoresist can be replicated, using a thermoplastic (see Section 9.2). Multiple copies of holographic optical elements (see Section 12.4) can also be made.

6.5 Photopolymers

Several organic materials can be activated by a photosensitizer to produce refractive index changes, due to photopolymerization, when exposed to light [Booth, 1977]. A commercial photopolymer is also available coated on a polyester film base (DuPont OmniDex) that can be used to produce volume phase holograms with high diffraction efficiency [Smothers *et al.*, 1990].

The film is supplied with a polyester cover sheet laminated on to the tacky photopolymer layer. Since the exposure required is around 300 J$/$m^2 at 514 nm, holograms are recorded by contact copying a master hologram (see Section 9.1). Close contact is ensured by removing the cover sheet and laminating the tacky film to the master hologram. The film is then exposed to UV light to cure the photopolymer, after which it can be separated from the master hologram. Finally the film is baked at 100–120 °C for 1–2 hours to obtain increased index modulation [Smothers *et al.*, 1990; Weber *et al.*, 1990].

This procedure yields volume reflection holograms that reconstruct an image at almost the same wavelength as that used to record them. If necessary, the reconstruction can be shifted to a longer wavelength by laminating the holographic recording film, after exposure to UV light, to a color tuning film and baking the sandwich [Zager & Weber, 1991].

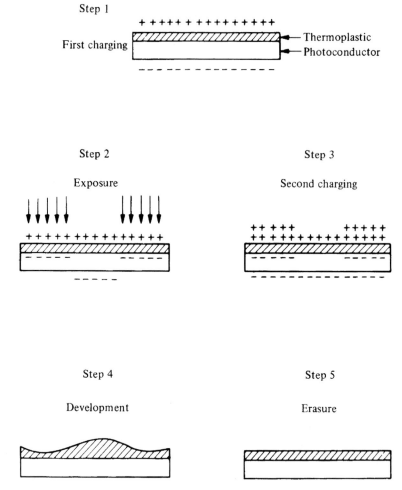

Fig. 6.4.   Record–erase cycle for a photothermoplastic [Lin and Beauchamp, 1970].

6.6 Photothermoplastics (PTP)

A hologram can be recorded in a multilayer structure consisting, as shown in fig. 6.4, of a glass or Mylar substrate coated with a thin, transparent, conducting layer of indium oxide, a photoconductor, and a thermoplastic [Lin & Beauchamp, 1970; Urbach, 1977].

The film is initially sensitized in darkness by applying a uniform electric charge to the top surface. On exposure and recharging, a spatially varying electrostatic field is created. The thermoplastic is then heated briefly, so that it becomes soft enough to be deformed by this field, and cooled to fix the variations in thickness.

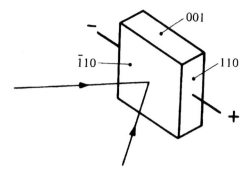

Fig. 6.5. Hologram recording configuration for BSO [Huignard, 1981].

Photothermoplastics have a reasonably high sensitivity and yield a thin phase hologram with good diffraction efficiency. They have the advantage that they can be processed rapidly *in situ*; in addition, if they are produced on a glass substrate, the hologram can be erased by heating the substrate, and the material reused.

6.7 Photorefractive crystals

When a photorefractive crystal is exposed to a spatially varying light pattern, electrons are liberated in the illuminated areas. These electrons migrate to adjacent dark regions and are trapped there. The spatially varying electric field produced by this space-charge pattern modulates the refractive index through the electro-optic effect, producing the equivalent of a phase grating. The space-charge pattern can be erased by uniformly illuminating the crystal, after which another recording can be made.

The photorefractive crystals most commonly used for recording holograms are Fe-doped $LiNbO_3$ and $Bi_{12}SiO_{20}$ (BSO),which has a higher sensitivity. The best results are obtained with BSO with the recording configuration shown in fig. 6.5, in which an electric field is applied at right angles to the hologram fringes [Huignard & Micheron, 1976; Huignard, 1981].

Typical recording–erasure curves for BSO at different applied fields, at an incident power density of 2.45 W/m^2 ($\lambda = 514$ nm) are presented in fig. 6.6. A maximum diffraction efficiency of 0.25 can be obtained with a field of 900 V/mm and the hologram can be stored in darkness for about 30 h. Since readout is destructive, the reconstructed image is best recorded and stored for viewing.

Several interesting possibilities have been opened up by such photorefractive materials.

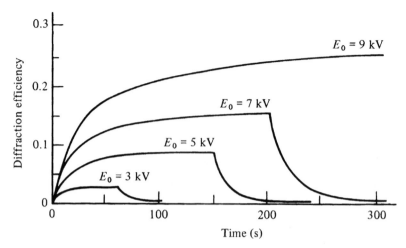

Fig. 6.6.   Record–erase cycles for BSO at different applied fields (sample size $10 \times 10 \times 10$ mm) [Huignard & Micheron, 1976].

References

Angell, D. K. (1987). Improved diffraction efficiency of silver halide (sensitized) gelatin. *Applied Optics*, **26**, 4692–702.

Bartolini, R. A. (1977). Photoresists. In *Holographic Recording Materials*, Topics in Applied Physics, vol.20, ed. H. M. Smith, pp. 209–27. Berlin: Springer-Verlag.

Belendez, A., Neipp, C., Flores, M. & Pascual, I. (1998). High-efficiency silver halide sensitized gelatin holograms with low absorption and scatter. *Journal of Modern Optics*, **45**, 1985–92.

Bjelkhagen, H. I. (1993). *Silver Halide Materials for Holography & Their Processing*. Berlin: Springer-Verlag.

Booth, B. L. (1977). Photopolymer laser recording materials. *Journal of Applied Photographic Engineering*, **3**, 24–30.

Chang, B. J. & Leonard, C. D. (1979). Dichromated gelatin for the fabrication of holographic optical elements. *Applied Optics*, **18**, 2407–17.

Fimia, A., Pascual, I. & Belendez, A. (1992). Optimized spatial frequency response in silver halide sensitized gelatin. *Applied Optics*, **31**, 4625–7.

Hariharan, P. (1986). Silver-halide sensitized gelatin holograms: mechanism of hologram formation. *Applied Optics*, **25**, 2040–2.

Hariharan, P. (1990). Basic processes involved in the production of bleached holograms. *Journal of Photographic Science*, **38**, 76–81.

Hariharan, P. & Chidley, C. M. (1988). Rehalogenating bleaches for photographic phase holograms. 2: Spatial frequency effects. *Applied Optics*, **27**, 3852–4.

Hariharan, P. & Chidley, C. M. (1989). Bleached reflection holograms: a study of color shifts due to processing. *Applied Optics*, **28**, 422–4.

Hariharan, P., Ramanathan, C. S. & Kaushik, G. S. (1971). Simplified processing technique for photographic phase holograms. *Optics Communications*, **3**, 246–7.

Huignard, J. P. (1981). Phase conjugation, real time holography and degenerate four-wave mixing in photoreactive BSO crystals. In *Current Trends in Optics*, ed. F. T. Arecchi & F. R. Aussenegg, pp. 150–60. London: Taylor & Francis.

Huignard, J. P. & Micheron, F. (1976). High sensitivity read-write volume holographic storage in $Bi_{12}SiO_{20}$ and $Bi_{12}GeO_{20}$ crystals. *Applied Physics Letters*, **29**, 591–3.

Jeong, M. H., Song, J. B. & Lee, I. W. (1991). Simplified processing method of dichromated gelatin holographic recording material. *Applied Optics*, **30**, 4172–3.

Lin, L. H. & Beauchamp, H. L. (1970). Write-read-erase in situ optical memory using thermoplastic holograms. *Applied Optics*, **9**, 2088–92.

Pennington, K. S., Harper, J. S. & Laming, F. P. (1971). New phototechnology suitable for recording phase holograms and similar information in hardened gelatine. *Applied Physics Letters*, **18**, 80–4.

Phillips, N. J., Ward, A. A., Cullen, R. & Porter, D. (1980). Advances in holographic bleaches. *Photographic Science & Engineering*, **24**, 120–4.

Smith, H. M. (1968). Photographic relief images. *Journal of the Optical Society of America*, **58**, 533–9.

Smith, H. M., ed. (1977). *Holographic Recording Materials*. Berlin: Springer-Verlag.

Smothers, W. K., Monroe, B. M., Weber, A. M. & Keys, D. E. (1990). Photopolymers for holography. In *Practical Holography IV*, Proceedings of the SPIE, vol. 1212, ed. S. A. Benton, pp. 20–9, Bellingham: SPIE.

Urbach, J. C. (1977). Thermoplastic hologram recording. In *Holographic Recording Materials*, Topics in Applied Physics, vol. 20, ed. H. M. Smith, pp. 161–207. Berlin: Springer-Verlag.

Weber, A. M., Smothers, W. K., Trout, T. J. & Mickish, D. J. (1990). Hologram recording in Du Pont's new photopolymer materials. In *Practical Holography IV*, Proceedings of the SPIE, vol. 1212, ed. S. A. Benton, pp. 30–9, Bellingham: SPIE.

Zager, S. A. & Weber, A. M. (1991). Display holograms in Du Pont's OmniDex films. In *Practical Holography V*, Proceedings of the SPIE, vol. 1461, ed. S. A. Benton, pp. 58–67, Bellingham: SPIE.

Problems

6.1. A hologram is to be recorded with a He–Ne laser on a BB-640 plate. The illumination level in the hologram plane due to the object beam is $0.003 \text{ W}/\text{m}^2$, and that due to the reference beam is $0.009 \text{ W}/\text{m}^2$. The plate is to be processed to produce a phase hologram. What is the exposure time required?

To obtain good diffraction efficiency after bleaching, the exposure should result in an optical density of at least 2.5 after development. From the published data for BB-640 plates, this would correspond to an exposure of approximately $1.5 \text{ J}/\text{m}^2$.

Accordingly, the exposure time required would be

$$T \approx 1.5 / (0.009 + 0.003)$$
$$\approx 125 \text{ seconds.} \tag{6.1}$$

The exposure time yielding the maximum diffraction efficiency can be selected from trials with exposures ranging from 100 seconds to 200 seconds.

7

Display holograms

7.1 Transmission holograms

A typical optical system for recording transmission holograms is shown in fig.7.1.

Making a hologram involves recording a two-beam interference pattern. Any change in the phase difference between the two beams during the exposure results in a movement of the fringes and reduces modulation in the hologram. Accordingly, to avoid mechanical disturbances, all the optical components, as well as the object and the photographic film or plate, should be mounted on a rigid surface resting on shock absorbers. The system is adjusted to have a low natural frequency of vibration (<1 Hz). A concrete or granite slab resting on inflated scooter inner tubes can be used, but most experimenters prefer a

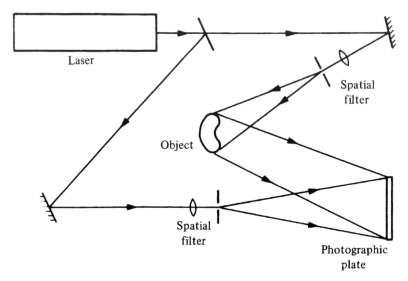

Fig. 7.1.   Optical arrangement used to record a transmission hologram.

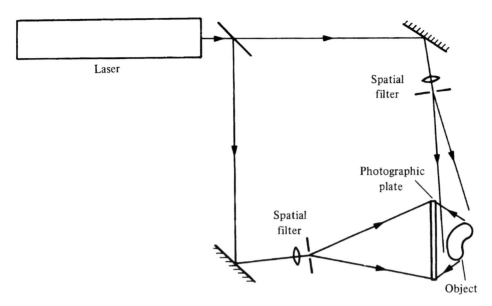

Fig. 7.2.   Optical arrangement used to record a reflection hologram.

honeycomb optical table, with a steel top, supported on pneumatic legs. The steel top has the advantage that optical components can be bolted down to its surface or mounted on magnetic bases. The effects of air currents and acoustic disturbances can be minimized by enclosing the working area with heavy curtains. An accurate power meter should be used to measure the intensities of the object and reference beams in the hologram plane, in order to set the ratio of their intensities to a suitable value and calculate the required exposure.

Where very long exposures have to be made, residual disturbances can be eliminated by a feedback system which stabilizes the optical path difference between the beams [Neumann & Rose, 1967]. Any motion of the interference fringes in the hologram plane is picked up by a photodetector, and the variations in its output are amplified and applied to a piezoelectric translator (PZT) which controls the position of one of the mirrors in the beam path.

7.2 Reflection holograms

A typical optical system that can be used for recording reflection holograms is shown schematically in fig. 7.2.

As can be seen, the object and reference waves are incident on the photographic emulsion from opposite sides. Since the thickness of the photographic emulsion is typically between 6 μm and 15 μm, the interference fringes are recorded as layers within it, about half a wavelength apart. The hologram exposure should

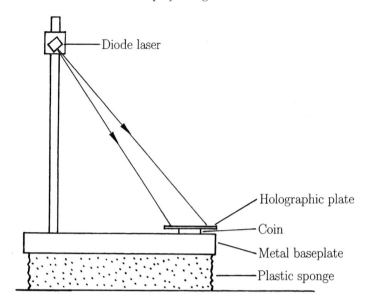

Fig. 7.3.   Recording arrangement for reflection holograms using a diode laser.

be adjusted to give a density greater than 2.0, after development. The hologram is then processed in a rehalogenating bleach to control emulsion shrinkage and the resulting wavelength shift [Hariharan & Chidley, 1989].

Reflection holograms of objects with limited depth can be made with an arrangement similar to that used originally by Denisyuk [1965], in which the portion of the reference beam transmitted by the photographic plate illuminates the object. A simple overhead configuration for recording such holograms, using a diode laser, has been described by Koch & Petros [1998].

Because of the small size of the diode laser, the entire optical system can be put together on a $20 \times 25 \times 1$ cm metal baseplate, which, as shown in fig. 7.3, is placed on a thick sheet of plastic sponge to provide vibration isolation. The diode laser (see Section 4.3) produces a clean beam whose divergence (typically, about $8° \times 25°$) makes it possible to illuminate the object through the holographic plate without any external optics. A hologram of a flat, specular reflecting object, such as a coin, can be recorded by placing a high resolution photographic plate directly on the object.

7.3 Full-view holograms

A drawback of conventional holograms is the limited angle over which they can be viewed. Holograms that give a full 360° view of an object can be recorded using either four plates, or a cylinder of film, surrounding the object.

A very simple optical system for this purpose [Jeong, 1967] is shown in

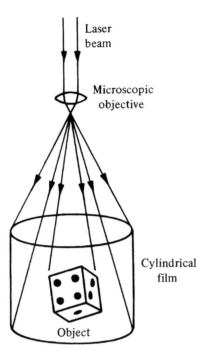

Fig. 7.4. Simple optical arrangement for making a 360° hologram [Jeong, 1967].

fig. 7.4. The object is placed at the center of a glass cylinder which has a strip of photographic film taped to its inner surface with the emulsion side facing inwards, and the expanded laser beam is incident on the object from above. The central portion of the expanded laser beam illuminates the object, while the outer portions, which fall directly on the film, constitute the reference beam.

To view the reconstructed image, the processed film is replaced in its original position and illuminated with the same laser beam.

7.4 Rainbow holograms

The rainbow hologram is a transmission hologram which reconstructs a bright, sharp, monochromatic image when illuminated with white light [Benton, 1977].

As shown schematically in fig. 7.5(*a*), the first step in making a rainbow hologram is to record a conventional transmission hologram of the object.

When this primary hologram (H$_1$) is illuminated, as shown in fig. 7.5(*b*), by the conjugate of the original reference wave, it reconstructs the conjugate of the original object wave and produces a real image of the object with unit magnification. A horizontal slit is then placed over H$_1$, as shown in figs. 7.5(*c*) and 7.5(*d*), and a second hologram (H$_2$) is recorded of the real image produced by H$_1$. The reference beam for H$_2$ is a convergent beam inclined in the vertical plane, and the photographic plate used to record H_2 is placed so that the real image formed by H_1 straddles it.

When H_2 is illuminated with the conjugate of the reference beam used to make it, it forms an orthoscopic image of the object straddling the plane of the hologram, as shown in fig. 7.6(*a*). In addition, it also forms a real image of the slit placed across H_1. All the light diffracted by the hologram passes through this slit pupil, so that a very bright image is seen from this position. Since the observer can move his head from side to side, horizontal parallax is retained. However, the image disappears if the observer's eyes move outside this slit pupil, so that vertical parallax is eliminated.

With a white light source, the slit image is dispersed in the vertical plane, as shown in fig. 7.6(*b*), to form a continuous spectrum. An observer whose eyes are positioned at any part of this spectrum then sees a sharp, three-dimensional image of the object in the corresponding color.

Figure 7.7 shows an optical system that permits both steps of the recording process to be carried out with a minimum of adjustments.

In this arrangement, the plane of the figure corresponds to the vertical plane in the final viewing geometry. A collimated reference beam is used to record the primary hologram (H_1), so that it is necessary only to turn H_1 through 180° about an axis normal to the plane of the figure and replace it in the holder, for an undistorted real image to be projected into the space in front of H_1. Vertical parallax is eliminated by a slit a few millimeters wide placed over H_1 with its long dimension normal to the plane of the figure. This orientation of the slit corresponds to the horizontal in the final viewing geometry. A convergent reference beam is used to record the final hologram (H_2), which, after processing, is reversed for viewing. When H_2 is illuminated with a divergent beam from a point source of white light, an orthoscopic image of the object is formed, and a dispersed real image of the slit is projected into the viewing space.

7.4.1 *Image blur*

The image reconstructed by a rainbow hologram is free from speckle, because it is illuminated with incoherent light, but not from blur [Wyant, 1977].

One cause of image blur is the finite wavelength spread in the image. If we consider a rainbow hologram made with the optical system shown schematically in fig. 7.8, the wavelength spread observed when the rainbow hologram is illuminated with white light is

$$\Delta\lambda = \left(\frac{\lambda}{\sin\theta}\right)\left(\frac{b+a}{D}\right), \tag{7.1}$$

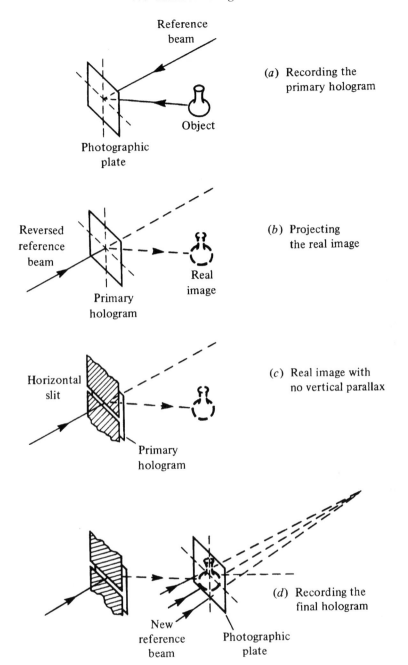

Fig. 7.5. Steps involved in the production of a rainbow hologram.

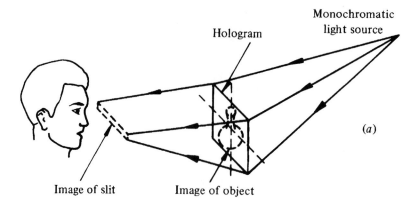

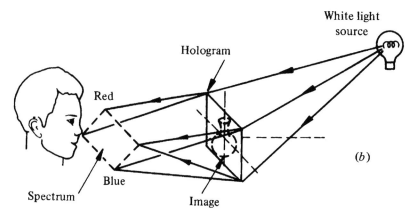

Fig. 7.6. Image reconstruction by a rainbow hologram (*a*) with monochromatic light, and (*b*) with white light.

where λ is the mean wavelength of the reconstructed image, θ is the angle made by the reference beam with the axis, b is the width of the slit, a is the diameter of the pupil of the eye and D is the distance from the primary hologram to the final rainbow hologram.

The image blur due to this wavelength spread is then

$$\Delta y_{\Delta\lambda} = z_0 \left(\frac{\Delta\lambda}{\lambda}\right) \sin\theta,$$

$$= z_0 \left(\frac{a+b}{D}\right), \tag{7.2}$$

where z_0 is the distance of the image from the hologram.

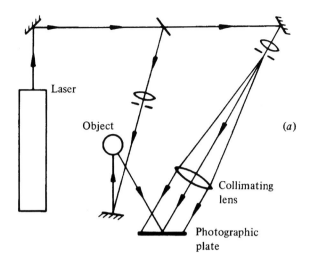

(a)

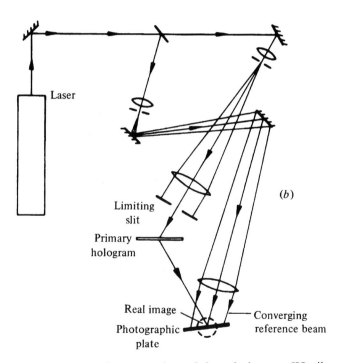

(b)

Fig. 7.7. Optical system used to record a rainbow hologram [Hariharan, Steel & Hegedus, 1977].

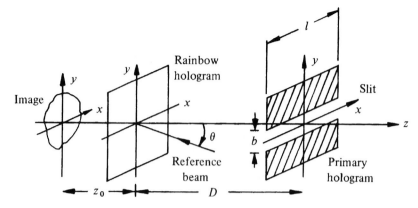

Fig. 7.8.   Analysis of image blur in a rainbow hologram [Wyant, 1977].

Another cause of image blur is the finite size of the source used to illuminate the hologram. If the source has an angular spread ψ_s, as viewed from the hologram, the resultant image blur is

$$\Delta y_s = z_0 \psi_s. \tag{7.3}$$

For the blur due to source size not to exceed the blur due to the wavelength spread, the size of the source must satisfy the condition

$$\psi_s \leq \left(\frac{a+b}{D}\right). \tag{7.4}$$

A final cause of image blur is diffraction at the slit; this can be neglected unless the width of the slit is very small.

7.5 Holographic stereograms

It is also possible to synthesize a hologram that reconstructs an acceptable three-dimensional image from a series of two-dimensional views of a subject from different angles [McCrickerd & George, 1968]. To produce such a holographic stereogram, a series of photographs of the subject is taken from equally spaced positions along a horizontal line. Alternatively, the subject is placed on a slowly rotating turntable, and a movie camera is used to make a record of a 120° or 360° rotation. Typically, three movie frames are recorded for each degree of rotation [Benton, 1975].

The optical system used to produce a white-light holographic stereogram from such a movie film is shown schematically in fig. 7.9. Each frame is imaged in the vertical plane on the film used to record the hologram. However, in the horizontal plane, the cylindrical lens brings the image to a line focus. A contiguous

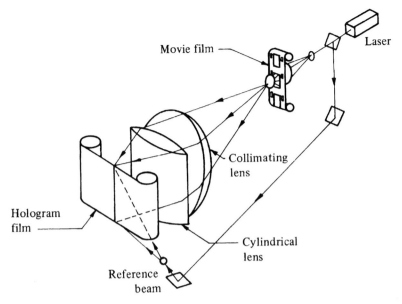

Fig. 7.9. Optical system used to produce a white-light holographic stereogram [Huff & Fusek, 1980].

sequence of vertical strip holograms is recorded of successive movie frames, covering the full range of views of the original subject, using a reference beam incident at an appropriate angle, either from above or from below. With a rotating subject, the processed film is formed into a cylinder for viewing.

When the holographic stereogram is illuminated with white light, the viewer sees a monochromatic three-dimensional image. This image changes color, as with a rainbow hologram, when the observer moves his head up or down. The image lacks vertical parallax, but it exhibits horizontal parallax over the range of angles covered by the original photographs.

Because of the difficulties involved in scaling up the system shown in fig. 7.9, large white-light holographic stereograms (up to 2 m × 1 m) are made by a two-step process [Newswanger & Outwater, 1985].

The obvious advantage of this technique, over recording a hologram directly, is that white light can be used to illuminate the subject in the first stage, so that holographic stereograms can be made of living subjects as well as large scenes. Some subject movement can also be displayed without destroying the stereoscopic effect.

7.6 Holographic movies

Moving three-dimensional images can be produced by presenting a sequence of holograms. With Fourier transform holograms (see Section 1.4), the reconstructed image can be viewed through the film by a single observer as the film is moved continuously. This limitation is acceptable for technical studies [Heflinger, Stewart & Booth, 1978; Smigielski, Fagot & Albe, 1985] but is a serious drawback for entertainment.

This problem was overcome [Komar, 1977; Komar and Serov, 1989] by using a lens with a diameter of 200 mm to record a series of image holograms of the scene on 70 mm film. The reconstructed image was projected with an identical lens on to a special holographic screen, equivalent to several superimposed concave mirrors. Each of these holographic mirrors then formed a real image of the projection lens in front of a spectator so that, when he looked through this pupil, he saw a full-size three-dimensional image.

Two artistic movies using holographic stereograms have also been produced by Alexander [Lucie-Smith, 1992].

References

Benton, S. A. (1975). Holographic displays – a review. *Optical Engineering*, **14**, 402–7.

Benton, S. A. (1977). White light transmission/reflection holographic imaging. In *Applications of Holography & Optical Data Processing*, ed. E. Marom, A. A. Friesem & E. Wiener-Avnear, pp. 401–9. Oxford: The Pergamon Press.

Denisyuk, Yu. N. (1965). On the reproduction of the optical properties of an object by the wave field of its scattered radiation. II. *Optics & Spectroscopy*, **18**, 152–7.

Hariharan, P. & Chidley, C. M. (1989). Bleached reflection holograms: a study of color shifts due to processing. *Applied Optics*, **28**, 422–4.

Hariharan, P., Steel, W. H. & Hegedus, Z. S. (1977). Multicolor holographic imaging with a white light source. *Optics Letters*, **1**, 8–9.

Heflinger, L. O., Stewart, G. L. & Booth, C. R. (1978). Holographic motion pictures of microscopic plankton. *Applied Optics*, **17**, 951–4.

Huff, L. & Fusek, R. L. (1980). Color holographic stereograms. *Optical Engineering*, **19**, 691–5.

Jeong, T. H. (1967). Cylindrical holography and some proposed applications. *Journal of the Optical Society of America*, **57**, 1396–8.

Koch, G. J. & Petros, M. (1998). A simple overhead Denisyuk configuration for making reflection holograms with a diode laser. *American Journal of Physics*, **66**, 933–4.

Komar, V. G. (1977). Progress on the holographic movie process in the USSR. In *Three-Dimensional Imaging*, Proceedings of the SPIE, vol. 120, ed. S. A. Benton, pp. 127–4. Redondo Beach: SPIE.

Komar, V. G. & Serov, O. B. (1989). Works on the holographic cinematography in the USSR. In *Holography '89*, Proceedings of the SPIE, vol. 1183, eds. Yu. N. Denisyuk & T. H. Jeong, pp. 170–82. Bellingham: SPIE.

Lucie-Smith, E. (1992). *Alexander*. London: Art Books International.

McCrickerd, J. T. & George, N. (1968). Holographic stereogram from sequential component photographs. *Applied Physics Letters*, **12**, 10–2.

Neumann, D. B. & Rose, H. W. (1967). Improvement of recorded holographic fringes by feedback control. *Applied Optics*, **6**, 1097–104.

Newswanger, C. & Outwater, C. (1985). Large format holographic stereograms and their applications. In *Applications of Holography*, Proceedings of the SPIE, vol. 523, ed. L. Huff, pp. 26–32. Bellingham: SPIE.

Smigielski, P., Fagot, H. & Albe, F. (1985). Progress in holographic cinematography. In *Progress in Holographic Applications*, Proceedings of the SPIE, vol. 600, ed. J. Ebbeni, pp. 186–93. Bellingham: SPIE.

Wyant, J. C. (1977). Image blur for rainbow holograms. *Optics Letters*, **1**, 130–2.

Problems

7.1 A rainbow hologram is recorded with an optical system similar to that shown in fig. 7.8, in which $\theta = 45°$, $D = 300$ mm and $b = 3$ mm. If the diameter of the pupil of the observer's eye is 3 mm and the image is reconstructed at a distance $z_0 = 50$ mm from the hologram, what is the image blur due to the wavelength spread? What is the minimum distance from the hologram at which a light source with a diameter of 10 mm should be placed for the image blur due to source size not to exceed that due to the wavelength spread?

From (7.2), the image blur due to the wavelength spread is

$$\Delta y_{\Delta\lambda} = z_0 \left(\frac{a+b}{D} \right)$$

$$= 50 \left(\frac{3+3}{300} \right) = 1 \text{ mm}. \tag{7.5}$$

For the image blur due to source size not to exceed that due to the wavelength spread, the angular extent of the source, viewed from the hologram, should satisfy (7.4). We have

$$\psi_s \leq \frac{a+b}{D}$$

$$\leq \frac{3+3}{300} \leq 0.02 \text{ radian}. \tag{7.6}$$

A 10 mm diameter source should be placed at a minimum distance of 500 mm from the hologram.

8

Multicolor images

In principle, a multicolor image can be produced by a hologram recorded with three suitably chosen wavelengths, when it is illuminated once again with these wavelengths. However, a problem is that each hologram diffracts, in addition to the wavelength used to record it, the other two wavelengths as well. The cross-talk images produced in this fashion overlap with, and degrade, the desired multicolored image. This problem has been overcome, and several methods are now available to produce multicolor images [Hariharan, 1983].

8.1 Multicolor reflection holograms

The first technique employed to eliminate cross-talk made use of the high wavelength selectivity of volume reflection holograms. If such a hologram is recorded with three wavelengths, one set of fringe planes is produced for each wavelength. When the hologram is illuminated with white light, each set of fringe planes diffracts a narrow band of wavelengths centered on the original wavelength used to record it, giving a multicolor image free from cross-talk [Upatnieks, Marks & Federowicz, 1966].

Higher diffraction efficiency can be obtained by superimposing three bleached volume reflection holograms recorded on two plates, one with optimum characteristics for the red, and the other with optimum characteristics for the green and blue. Brighter images can also be obtained if the final holograms are produced using real images of the object projected by primary holograms whose aperture is limited by a suitably shaped stop [Hariharan, 1980a].

8.2 Multicolor rainbow holograms

Another method of producing multicolor images makes use of superimposed rainbow holograms [Hariharan, Steel & Hegedus, 1977]. In this technique, three primary holograms are made with red, green and blue laser light. These

72

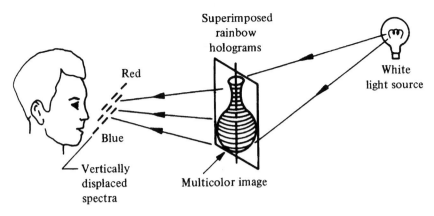

Fig. 8.1.   Reconstruction of a multicolor image by superimposed rainbow holograms [Hariharan, 1983].

primary holograms are then used with the same laser sources to make a set of three rainbow holograms, which are then superimposed. When this multiplexed hologram is illuminated with a white light source, it reconstructs three images of the object. In addition, as shown in fig. 8.1, three dispersed images of the slit are formed in the viewing space. Since the corresponding three spectra are displaced by appropriate amounts with respect to each other, an observer, viewing the hologram from the original position of the slit, sees three superimposed images of the object reconstructed in the colors with which the primary holograms were made.

Rainbow holograms can be used very effectively in multicolor displays, since the reconstructed images are very bright and exhibit high color saturation, and are also free from cross-talk.

8.3 Light sources

The most commonly used lasers for color holography are the He–Ne laser which provides an output in the red ($\lambda = 633$ nm), and the Ar$^+$ laser which provides outputs in the green ($\lambda = 514$ nm) and the blue ($\lambda = 488$ nm). The range of wavelengths that can be reconstructed with these three wavelengths as primaries can be determined by means of the C.I.E. chromaticity diagram. As shown in fig. 8.2, points representing monochromatic light of different wavelengths are located on a horseshoe-shaped curve known as the spectrum locus; all other colors lie within this boundary. New colors obtained by mixing light of two wavelengths, such as 633 nm and 514 nm, lie on the straight line AB joining these primaries. If light with a wavelength of 488 nm is also used, any color within the triangle ABC can be obtained.

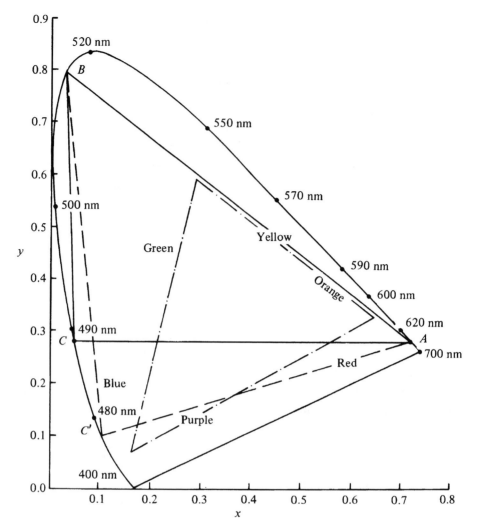

Fig. 8.2.　C.I.E. chromaticity diagram. The triangle ABC shows the range of hues that can be produced by a hologram illuminated with primary wavelengths of 633 nm, 514 nm and 488 nm, while the broken lines show the extended range obtained with a blue primary at 477 nm. The chain lines enclose the range of hues that can be reproduced by a typical color-television display [Hariharan, 1983].

A wider range of colors can be obtained, at some sacrifice of power, by using the blue output from an Ar⁺ laser at a wavelength of 477 nm, or a He–Cd laser at 422 nm.

8.4 Pseudocolor images

The color information in a hologram is encoded in the spatial frequencies of the carrier fringes. It is therefore possible to generate different carrier fringe frequencies and, consequently, images of different colors, even with a single laser wavelength, by different means.

8.4.1 Pseudocolor rainbow holograms

One way to produce a multicolor image is to record three superimposed rainbow holograms with different reference beam angles [Tamura, 1978]. Alternatively the position of the limiting slit can be changed between exposures.

A problem is that the images reconstructed in a different color from that used to record the hologram are displaced with respect to it. The displacement in the vertical plane is

$$\Delta y = z_0 \left(\frac{2 \Delta \lambda}{\lambda} \right) \tan^3 \theta, \qquad (8.1)$$

while the longitudinal displacement is

$$\Delta z = z_0 \left(\frac{2 \Delta \lambda}{\lambda} \right), \qquad (8.2)$$

where z_0 is the distance of the image from the hologram, θ is the interbeam angle in the recording system (see fig. 7.8), λ is the recording wavelength and $(\lambda + \Delta \lambda)$ is the wavelength at which the image is reconstructed. These displacements can be tolerated if the image is formed close to the hologram.

8.4.2 Pseudocolor reflection holograms

With volume reflection holograms, the color of the reconstructed image is affected by changes in the thickness of the recording medium, and these changes can be controlled and used to produce pseudocolor images even with a He–Ne laser [Hariharan, 1980b].

The red component hologram is recorded first, on a plate exposed with the emulsion side towards the reference beam. This plate is processed using a rehalogenating bleach to minimize emulsion shrinkage. The green and blue images are then recorded on another plate exposed with the emulsion side towards the object beam. After the green component is exposed, the emulsion is soaked in a 3 percent solution of triethanolamine, to swell it, and dried in darkness. The blue component is then exposed. Normal bleach processing eliminates the triethanolamine and produces the usual shrinkage. As a result, the first exposure yields a green image, while the second produces a blue image.

After drying, the plates are cemented together, with the emulsions in contact, and viewed with the hologram reconstructing the blue and green images in front.

Very high quality pseudocolor images can be produced from contact copies (see Section 9.1) of three master holograms made on DuPont photopolymer Holographic Recording Film with an Ar^+ laser (476 nm). Two of the holograms are swollen during processing by baking in contact with DuPont Color Tuning Film, so that they reconstruct green and red images, before the three films are sandwiched together [Hubel & Klug, 1992].

References

Hariharan, P. (1980*a*). Improved techniques for multicolour reflection holograms. *Journal of Optics* (*Paris*), **11**, 53–5.

Hariharan, P. (1980*b*). Pseudocolour images with volume reflection holograms. *Optics Communications*, **35**, 42–4.

Hariharan, P. (1983). Colour Holography. In *Progress in Optics*, vol. 20, ed. E. Wolf. pp. 265–324. Amsterdam: North-Holland.

Hariharan, P., Steel, W. H. & Hegedus, Z. S. (1977). Multicolor holographic imaging with a white light source. *Optics Letters*, **1**, 8–9.

Hubel, P. & Klug, M. A. (1992). Color holography using multiple layers of DuPont photopolymer. In *Practical Holography V*, Proceedings of the SPIE, vol. 1667, ed. S. A. Benton, pp. 215–24. Bellingham: SPIE.

Tamura, P. N. (1978). Pseudocolor encoding of holographic images using a single wavelength. *Applied Optics*, **17**, 2532–6.

Upatnieks, J., Marks, J. & Federowicz, R. (1966). Color holograms for white light reconstruction. *Applied Physics Letters*, **8**, 286–7.

Problems

8.1. What is the theoretical improvement in diffraction efficiency possible in a multicolor rainbow hologram by recording the three component holograms on three separate plates instead of on a single plate?

If all the three holograms are recorded on a single photographic plate, the available dynamic range is divided between the three holograms (see Section 3.6). The diffraction efficiency of each component hologram is then only $(1/3)^2 = 1/9$ of that for a single hologram recorded on the same plate. The diffraction efficiency should therefore improve, theoretically, by a factor of 9 (the actual gain is about half of this, because of transmission losses).

8.2. A pseudocolor rainbow hologram is recorded with an Ar^+ laser ($\lambda = 514$ nm) with an interbeam angle of 30°. While recording the red and blue components, the slit is shifted so that these images are reconstructed at wavelengths of 630 nm and 450 nm respectively. If the lateral displacements of the red and blue images are not to exceed 1 mm, what is the maximum permissible depth of these images?

We consider, first, the displacement of the red image, since the wavelength shift for it is greater ($\Delta\lambda = 116$ nm). From (8.1), we have

$$\Delta y = z_0 \left(\frac{2\Delta\lambda}{\lambda}\right) \tan^3 \theta,$$

$$= z_0 \left(\frac{2 \times 116}{514}\right) \left(\frac{1}{\sqrt{3}}\right)^3$$
$$= 0.124 \times z_0. \tag{8.3}$$

Accordingly, for the lateral displacement of the edges of the red image to be less than 1 mm, they must lie within ± 8 mm of the hologram.

A similar calculation shows that the edges of the blue image (for which $\Delta\lambda = 64$ nm) must lie within ± 14 mm of the hologram.

9

Copying holograms

It is often necessary to make copies from a single original hologram.

One way is to illuminate the hologram with the conjugate of the reference wave used to make it. The wave reconstructed by the hologram can then be used with another reference wave to record a second-generation hologram. This technique has the advantage of great flexibility. It is possible to produce a copy that reconstructs an orthoscopic real image (see Section 2.2), as well as copies with improved diffraction efficiency. It is also possible to produce a number of reflection holograms from a transmission hologram of the object.

9.1 'Contact printing'

A simpler way to produce many identical copies of a hologram is to 'contact print' the original on to another photosensitive material [Harris, Sherman & Billings, 1966].

Since what is recorded on the copy material is actually the interference pattern formed by the light diffracted by the hologram and the light transmitted by it, the coherence of the illumination must be adequate to produce interference fringes of high visibility. If, as shown in fig. 9.1, the primary hologram (H_1) diffracts light at an angle θ, interference takes place between rays originally separated by a distance Δx where

$$\Delta x = \Delta z \tan \theta, \tag{9.1}$$

and the optical path difference between the beams is

$$\Delta l = \Delta x \sin \theta$$
$$= \Delta z \tan \theta \sin \theta, \tag{9.2}$$

so that the coherence requirements for copying are much less stringent than for recording a hologram.

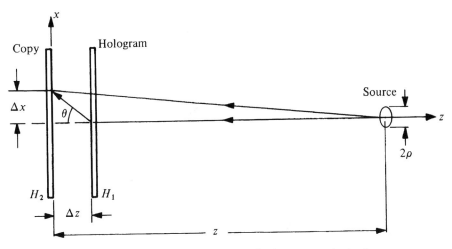

Fig. 9.1. Copying a hologram by 'contact printing'.

To obtain a copy with good diffraction efficiency, the exposure and processing conditions for the primary hologram must be chosen so that the amplitudes
of the transmitted wave and the diffracted wave are comparable. In addition,
for the best results, the direction, curvature and wavelength of the original reference wavefront must be maintained in the copying system, and an index-
matching fluid should be used between the hologram and the copy film to
eliminate spurious interference fringes formed by reflection between the surfaces.

9.2 Embossed holograms

Holograms recorded on a photoresist can be copied by embossing [Iwata &
Tsujiuchi, 1974].

The first step in the embossing process is to make a stamper by electrodepo-
sition of nickel on the relief image recorded on the photoresist [Iwata &
Ohnuma, 1985]. When the nickel layer is thick enough, it is separated from the
master hologram and mounted on a metal backing plate.

Figure 9.2 is a cross section of the material used to make embossed copies.
It consists of a polyester base film, a resin separation layer and a thermoplastic film (the hologram layer).

The embossing process can be carried out with a simple heated press, as
shown in fig. 9.3. The bottom layer of the duplicating film (the thermoplastic
layer) is heated above its softening point and pressed against the stamper. The
thermoplastic layer (the hologram layer) then takes up the shape of the
stamper and retains this shape when it is cooled and removed from the press.

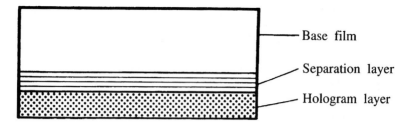

Fig. 9.2.   Cross section of a film used to make embossed copies of holograms.

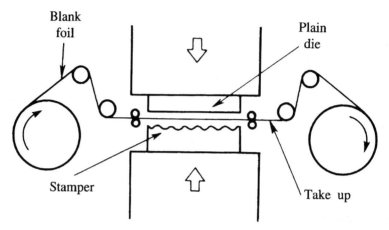

Fig. 9.3.   Embossing press used to make copies of holograms.

A roll press can be used for mass replication of large format embossed holograms [Burns, 1985].

Embossed holograms can be transferred to an opaque surface, such as the cover of a book, by transcription. To permit viewing by reflected light, the transcription foil contains two more layers, a reflecting layer of aluminum deposited in vacuum on the hologram recording layer and an adhesive layer. When the transcription foil is placed on the substrate to which the hologram is to be transfered and pressed with a heated die, the bottom adhesive layer melts and sticks to the substrate. After it cools, the base film can be lifted off, leaving the other layers, including the hologram, attached to the substrate.

Embossed holograms are now used widely as a security feature on credit cards and quality merchandise [Fagan, 1990].

References

Burns, J. R. (1985). Large-format embossed holograms. In *Applications of Holography*. Proceedings of the SPIE, vol. 523, ed. L. Huff, pp. 7–14. Bellingham, SPIE.

Fagan, W. F., ed. (1990). *Optical Security & Anticounterfeiting Systems*, Proceedings of the SPIE, vol. 1210. Bellingham: SPIE.

Harris, Jr., F. S., Sherman, G. C. & Billings, B. H. (1966). Copying holograms. *Applied Optics*, **5**, 665–6.

Iwata, F. & Tsujiuchi, J. (1974). Characteristics of a photoresist hologram and its replica. *Applied Optics*, **13**, 1327–36.

Iwata, F. & Ohnuma, K. (1985). Brightness and contrast of a surface relief rainbow hologram for an embossing master. In *Applications of Holography*. Proceedings of the SPIE, vol. 523, ed. L. Huff, pp. 15–17. Bellingham, SPIE.

Problems

9.1. We would like to make copies of a hologram recorded with an interbeam angle of 30° by 'contact printing'. If we assume that the maximum separation between the hologram and the copy film is 100 μm, what is the lower limit on the coherence length of the light used to make the copy?

From (9.2), the contrast of the carrier fringes in the copy would drop to zero if the coherence length of the light is less than

$$\Delta l = \tan 30° \times \sin 30° \times 100 \times 10^{-6} \text{ m}$$
$$= 34 \text{ μm.} \tag{9.3}$$

For a negligible loss in contrast, the coherence length of the light used to make a copy should be much greater than this figure, say, 1 mm.

10

Computer-generated holograms

Computer-generated holograms can produce wavefronts with any prescribed amplitude and phase distribution and have, therefore, found many applications. The production of such holograms has been discussed by Lee [1978], Yaroslavskii and Merzlyakov [1980] and Dallas [1980], and essentially involves two steps.

The first step is to calculate the complex amplitude of the object wave at the hologram plane; for simplicity, this is usually taken to be the discrete Fourier transform (see Appendix B) of the complex amplitude at an $N \times N$ set of points in the object plane. The second step involves using the $N \times N$ computed values of the discrete Fourier transform to produce a transparency (the hologram) which reconstructs the object wave when it is suitably illuminated.

Two approaches have been followed for this purpose. In the first, which is analogous to off-axis holography, the complex amplitudes of a plane reference wave and the object wave, at each point in the hologram plane, are added, and the squared modulus of their sum is evaluated. These values are used to produce a transparency whose amplitude transmittance is real and positive everywhere.

An alternative is to produce a transparency that records both the amplitude and the phase of the object wave in the hologram plane. This transparency can be thought of as the superposition of two transparencies, one of constant thickness having a transmittance at each point proportional to the amplitude of the object wave, and the other with uniform transmittance but having thickness variations proportional to the phase of the object wave. Such a hologram has the advantage that it forms a single, on-axis image.

In either case, the output from the computer is used to control a plotter that produces a large scale version of the hologram. This master is photographically reduced to produce the required transparency.

10.1 Binary detour-phase holograms

Production of the hologram can be simplified considerably if its transmittance has only two levels – either zero or one. The best known hologram of this type is the binary detour-phase hologram [Brown & Lohmann, 1966, 1969]. To produce such a hologram, the output format covered by the plotter is divided into $N \times N$ cells, which correspond to the $N \times N$ coefficients of the discrete Fourier transform of the complex amplitude in the object plane. Each complex Fourier coefficient is then represented by a single transparent area within the corresponding cell, whose size is determined by the modulus of the Fourier coefficient, while its position within the cell represents the phase of the Fourier coefficient. A shift of the transparent area in any cell results in the light transmitted by it traveling by a longer or shorter path to the reconstructed image, hence the name of the method.

Figure 10.1(*a*) shows a binary detour-phase hologram of the letters ICO; fig. 10.1(*b*) shows the image produced by it. The first-order images are those above and below the central spot; the higher-order images are due to nonlinear effects.

To understand how this method of encoding the phase works, consider a rectangular opening $(a \times b)$ in an opaque sheet (the hologram) centered on the origin of coordinates, as shown in fig. 10.2, which is illuminated with a uniform plane wave of unit amplitude.

The complex amplitude $U(x_i, y_i)$ in the diffraction pattern formed in the far field is given by the Fourier transform of the transmitted amplitude and is

$$U(x_i, y_i) = ab \ \text{sinc}(by_i / \lambda z), \tag{10.1}$$

where $\text{sinc } x = (\sin \pi x) / \pi x$.

We now assume that the center of the rectangular opening is shifted to a point $(\Delta x_0, \Delta y_0)$, and the sheet is illuminated by a plane wave incident at an angle. If the complex amplitude of the incident wave at the sheet is $\exp[i(\alpha \Delta x_0 + \beta \Delta y_0)]$, the complex amplitude in the diffraction pattern becomes

$$U(x_i, y_i) = ab \ \text{sinc}(ax_i / \lambda z) \ \text{sinc}(by_i / \lambda z)$$

$$\times \exp\left[i\left(\alpha + \frac{2\pi x_i}{\lambda z}\right)\Delta x_0 + i\left(\beta + \frac{2\pi y_i}{\lambda z}\right)\Delta y_0\right],$$

$$= ab \ \text{sinc}(ax_i / \lambda z) \ \text{sinc}(by_i / \lambda z)$$
$$\times \exp[i(\alpha \Delta x_0 + \beta \Delta y_0)]$$

$$\times \exp\left[i\left(\frac{2\pi}{\lambda z}x_i\Delta x_0 + \frac{2\pi}{\lambda z}y_i\Delta y_0\right)\right]. \tag{10.2}$$

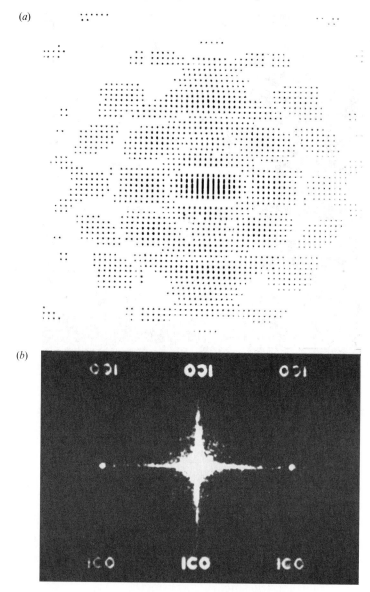

Fig. 10.1. Binary detour-phase hologram: (*a*) the hologram; (*b*) the reconstructed image [Lohmann & Paris, 1967].

If $ax_i \ll \lambda z$, $by_i \ll \lambda z$, (10.2) reduces to

$$U(x_i, y_i) = ab \, \exp[\mathrm{i}(\alpha \Delta x_0 + \beta \Delta y_0)]$$

$$\times \exp\left[\mathrm{i}\left(\frac{2\pi}{\lambda z} x_i \Delta x_0 + \frac{2\pi}{\lambda z} y_i \Delta y_0\right)\right]. \tag{10.3}$$

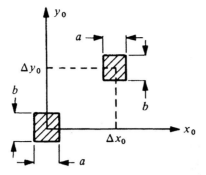

Fig. 10.2. Diffraction at a rectangular aperture.

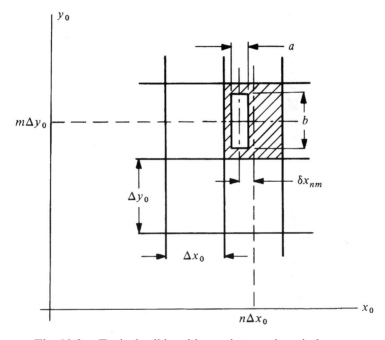

Fig. 10.3. Typical cell in a binary detour-phase hologram.

If, then, the computed complex amplitude of the object wave at a point $(n\Delta x_0, m\Delta y_0)$ in the hologram plane is

$$o(n\Delta x_0, m\Delta y_0) = |o(n\Delta x_0, m\Delta y_0)| \exp[i\phi(n\Delta x_0, m\Delta y_0)], \qquad (10.4)$$

its modulus and phase at this point can be encoded, as shown in fig. 10.3, by making the area of the opening in this cell equal to the modulus, so that

$$ab = |o(n\Delta x_0, m\Delta y_0)|, \qquad (10.5)$$

and displacing the center of the opening from the center of the cell by an amount δx_{nm} given by the relation

$$\delta x_{nm} = (\Delta x_0 / 2\pi)\, \phi(n\Delta x_0, m\Delta y_0). \tag{10.6}$$

To confirm the validity of this technique, we consider the complex amplitude in the far field due to this opening, which, from (10.3), is

$$U_{nm}(x_i, y_i) = |o(n\Delta x_0, m\Delta y_0)|$$

$$\times \exp[i\alpha(n\Delta x_0 + \delta x_{nm}) + i\beta m\Delta y_0]$$

$$\times \exp[(i2\pi / \lambda z)\,(n x_i\,\Delta x_0 + m y_i\,\Delta y_0 + \delta x_{nm})]. \tag{10.7}$$

The total diffracted amplitude in the far field, which is obtained by summing the complex amplitudes due to all the $N \times N$ openings is, therefore,

$$U(x_i, y_i) = \sum_{n=1}^{N} \sum_{m=1}^{N} |o(n\Delta x_0, m\Delta y_0)|\, \exp(i\alpha\delta x_{nm})$$

$$\times \exp[i(\alpha n\Delta x_0 + \beta m\Delta y_0)]$$
$$\times \exp[(i2\pi / \lambda z)(n x_i\Delta x_0 + m y_i\Delta y_0)]$$
$$\times \exp[(i2\pi / \lambda z)\delta x_{nm}]. \tag{10.8}$$

If the dimensions of the cells and the angle of illumination are chosen so that

$$\alpha\Delta x_0 = 2\pi, \tag{10.9}$$
$$\beta\Delta y_0 = 2\pi, \tag{10.10}$$
$$\delta x_{nm} \ll \lambda z, \tag{10.11}$$

(10.8) reduces to

$$U(x_i, y_i) = \sum_{n=1}^{N} \sum_{m=1}^{N} |o(n\Delta x_0, m\Delta y_0)|\, \exp[i\phi(n\Delta x_0, m\Delta y_0)]$$

$$\times \exp[(i2\pi / \lambda z)(n x_i\Delta x_0 + m y_i\Delta y_0)], \tag{10.12}$$

which is the discrete Fourier transform of the computed complex amplitude in the hologram plane, that is to say, the desired reconstructed image.

Binary detour-phase holograms have the advantage that it is possible to use a simple pen-and-ink plotter to prepare the binary master, and problems of linearity do not arise in the photographic reduction process. Their chief disadvantage is that they are very wasteful of plotter resolution, since the number of addressable plotter points in each cell must be large to minimize the noise due to quantization of the modulus and the phase of the Fourier coefficients.

Procedures for making such holograms for lecture demonstrations, using a desktop computer, have been described in several publications [McGregor, 1992]. The availability of high-resolution laser printers has eliminated the need for photographic reduction, and 75×75 element, detour-phase holograms can be produced directly on overhead transparency film [Walker, 1999].

10.2 Phase randomization

The Fourier transforms of the wavefronts corresponding to most simple objects have very large dynamic ranges, because the coefficients of the dc and low-frequency terms have much larger moduli than those of the high-frequency terms.

Where the phase of the final reconstructed image is not important, this problem can be minimized by multiplying the complex amplitudes at the original sampled object points by a random phase factor before calculating the Fourier transform [Lohmann & Paris, 1967]. This procedure is optically equivalent to placing a diffuser in front of the object transparency, and has the effect of making the magnitudes of the coefficients of the Fourier transforms much more uniform. However, as shown in fig. 10.4, the reconstructed image is then modulated by a speckle pattern.

Several coding techniques have been developed to reduce quantization errors and achieve a satisfactory compromise between smoothing the object spectrum and minimizing speckle [Bryngdahl & Wyrowski, 1990].

10.3 Three-dimensional objects

A three-dimensional object can be approximated by the sum of a number of equally spaced cross sections perpendicular to the z axis. However, problems can arise from distant parts of the object, which are normally hidden by surfaces in front, appearing in the image. It is then necessary, at each point on the hologram, to sum only contributions to the object wave arising from points on the object that can be seen from that point on the hologram.

The generation of a three-dimensional image involves a very large amount of computation. One way to reduce the computing time is to eliminate vertical parallax by performing a series of Fourier transforms corresponding to successive horizontal lines in the hologram [Leseberg, 1986]. Another, described by King, Noll and Berry [1970], uses a technique similar to that for making holographic stereograms (see Section 7.5). A computer is used to produce a series of views of the object, as seen from a number of angles in the horizontal plane, and these views are optically encoded as a series of vertical strip holograms on

(*a*)

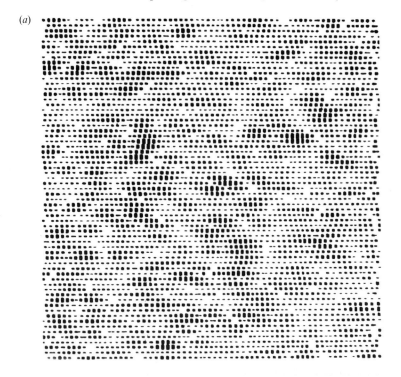

(*b*)

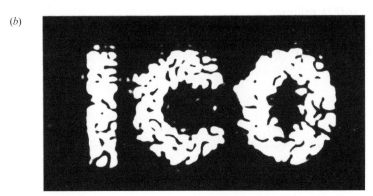

Fig. 10.4.   Object with a random phase: (*a*) the hologram; (*b*) the reconstructed image [Lohmann & Paris, 1967].

a single piece of film. Finally, the real image formed by this composite holo-gram, when it is illuminated by the conjugate reference beam, is used to produce an image hologram. Since this real image consists of a series of two-dimensional images which are located entirely in the plane of the final holo-gram, it can be illuminated with white light to reconstruct a bright, almost achromatic image.

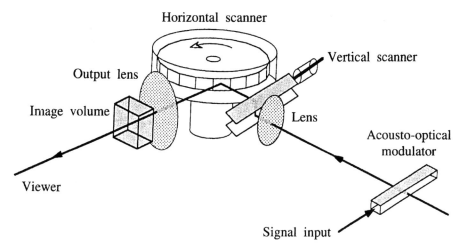

Fig. 10.5.   Schematic of a holographic video display system [St.-Hilaire *et al.*, 1990].

10.4 Holographic video imaging

A real-time holographic display of three-dimensional information is also possible with computer-generated holograms [St.-Hilaire *et al.*, 1990].

Data on a synthetic three-dimensional object are transferred to a computer containing 16 000 microprocessors in a massively parallel architecture. The video signal from a frame buffer is used to drive an acousto-optic modulator (AOM) in the display system shown in fig. 10.5. An expanded laser beam emerges from this AOM with a phase modulation across its width that is proportional to the input signal representing a section of one horizontal line of the hologram. A spinning polygonal mirror and a galvanometer scanner are used to multiplex these sections in the horizontal and vertical directions, respectively, to build up the holographic image. Images up to 130 mm × 170 mm and 200 mm deep have been produced by this system.

Multicolor images have also been generated with such a system by using a three-channel AOM and illuminating the three channels with red, green and blue lasers [St.-Hilaire *et al.*, 1992].

10.5 Optical testing

One of the main applications of computer-generated holograms is in optical testing, where they are used in interferometric tests of aspheric surfaces. In this application, a computer-generated hologram replaces an expensive null lens used to cancel the aberrations of the test wavefront [Wyant & Bennett, 1972].

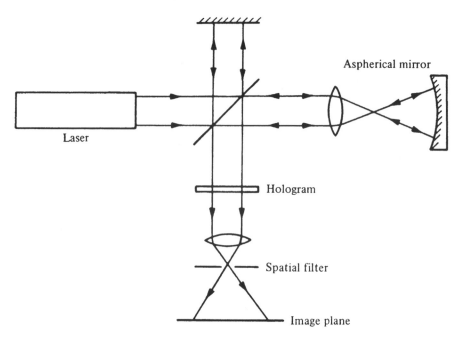

Fig. 10.6. Twyman–Green interferometer modified to use a computer-generated hologram to test an aspherical mirror [Wyant & Bennett, 1972].

The hologram is a representation of the interferogram that would be obtained if the wavefront from the desired aspheric surface were to interfere with a tilted plane wavefront. If, as shown in fig. 10.6, the test surface is imaged on the hologram, the superposition of the actual interference pattern and the hologram produces a moiré pattern showing the deviation of the test wavefront from the ideal computed wavefront.

The contrast of the interference pattern is improved by reimaging the hologram through a small aperture placed in the focal plane of the reimaging lens which passes only the transmitted wavefront from the mirror under test and the diffracted wavefront produced by illuminating the hologram with the plane reference wavefront. Typical fringe patterns obtained with an aspheric surface, with and without a computer-generated hologram, are shown in fig. 10.7.

To obtain good results, several sources of error must be kept in mind. One is the effect of quantization. If there are N resolvable points across the diameter of the hologram, any plotted point may be displaced from its position by $1/2N$ of the diameter. However, for the diffracted images not to overlap (see Appendix B), the hologram must have a carrier frequency $S_H = 3S_I$, where S_I is the highest spatial frequency (fringes per diameter) in the uncorrected fringe

(a)   (b)

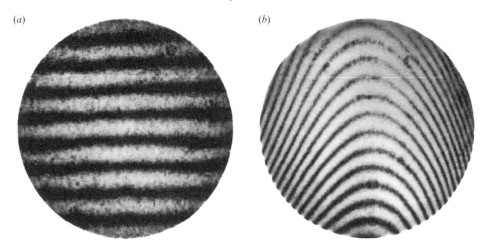

Fig. 10.7.   Interference pattern obtained with an aspheric mirror (*a*) with, and (*b*) without a computer-generated hologram [Wyant & Bennett, 1972].

pattern. As a result, the fringe frequency S_H in the hologram can vary from a minimum of $2S_I$ to a maximum of $4S_I$. With a fringe frequency $S_H = 4S_I$, an error in the fringe position of $1/2N$ would correspond to a wavefront error (expressed as a fraction of a fringe)

$$\Delta W = 2S_I / N. \qquad (10.13)$$

If the wavefront error is not to exceed (say) $\lambda/4$, the number of resolvable points across the diameter of the hologram must satisfy the condition

$$N > 4 \times 2S_I. \qquad (10.14)$$

Other sources of error are plotter distortion and errors in the size and positioning of the hologram.

Electron-beam recording on layers of photoresist coated on optically worked blanks now makes it possible to produce computer-generated holograms of very high quality which are used widely to test aspheric surfaces [Arnold, 1985, 1989].

References

Arnold, S. M. (1985). Electron-beam fabrication of computer-generated holograms. *Optical Engineering*, **24**, 803–7.

Arnold, S. M. (1989). How to test an asphere with a computer generated hologram. In *Holographic Optics: Optically & Computer Generated*, Proceedings of the SPIE, vol. 1052, eds. I. Cindrich & S. H. Lee, pp. 191–7. Bellingham: SPIE.

Brown, B. R. & Lohmann, A. W. (1966). Complex spatial filtering with binary masks. *Applied Optics*, **5**, 967–9.

Brown, B. R. & Lohmann, A. W. (1969). Computer generated binary holograms. *IBM Journal of Research & Development*, **13**, 160–7.

Bryngdahl, O. & Wyrowski, F. (1990). Digital holography – computer-generated holograms. In *Progress in Optics*, vol. 28, ed. E. Wolf, pp. 1–86. Amsterdam: North-Holland.

Dallas, W. J. (1980). Computer-generated holograms. In *The Computer in Optical Research*, Topics in Applied Physics, vol. 41, ed. B. R. Frieden, pp. 291–396. Berlin: Springer-Verlag.

King, M. C., Noll, A. M. & Berry, D. H. (1970). A new approach to computer-generated holography. *Applied Optics*, **9**, 471–5.

Lee, W. H. (1978). Computer-generated holograms: techniques and applications. In *Progress in Optics*, vol. 16, ed. E. Wolf, pp. 121–232. Amsterdam: North-Holland.

Leseberg, D. (1986). Computer-generated holograms: display using one-dimensional transforms. *Journal of the Optical Society of America. A*, **3**, 1846–51.

Lohmann, A. W. & Paris, D. P. (1967). Binary Fraunhofer holograms generated by computer. *Applied Optics*, **6**, 1739–48.

Macgregor, A. E. (1992). Computer generated holograms from dot matrix and laser printers. *American Journal of Physics*, **60**, 839–46.

St.-Hilaire, P., Benton, S. A., Lucente, M., Jepsen, M. L., Kollin, J., Yoshikawa, H. & Underkoffler, J. (1990). Electronic display system for computational holography. In *Practical Holography IV*, Proceedings of the SPIE, vol. 1212, ed. S. A. Benton, pp. 174–82. Bellingham: SPIE.

St.-Hilaire, P., Benton, S. A., Lucente, M. & Hubel, P. M. (1992). Color images with the MIT holographic display. In *Practical Holography VI*, Proceedings of the SPIE, vol. 1667, ed. S. A. Benton, pp. 73–84. Bellingham: SPIE.

Walker, T. G. (1999). Holography without photography. *American Journal of Physics*, **67**, 783–5.

Wyant, J. C. & Bennett, V. P. (1972). Using computer-generated holograms to test aspheric wavefronts. *Applied Optics*, **11**, 2833–9.

Yaroslavskii, L. P. & Merzlyakov, N. S. (1980). *Methods of Digital Holography*. New York: Consultants Bureau, Plenum Publishing Co.

Problems

10.1. A computer-generated hologram is to be used instead of a null lens to test an aspheric wavefront which has a maximum departure from the reference sphere of 19 wavelengths and a maximum slope error of 35 waves per radius. What is (*a*) the minimum carrier fringe frequency that can be used, and (*b*) the minimum number of resolvable points across the hologram for the residual wavefront errors to be less than $\lambda/8$?

In this case, the highest spatial frequency (fringes per diameter) in the uncorrected interference pattern is $S_I = 2 \times 35 = 70$ fringes per diameter.

For the diffracted images not to overlap, the hologram must have a carrier fringe frequency $S_H \geq 3S_I$, Accordingly, the minimum carrier fringe frequency is

$$S_H = 3 \times 70$$
$$= 210 \text{ fringes per diameter.} \tag{10.15}$$

For the wavefront error to be less than $\lambda/8$, it then follows, from (10.14), that the number of resolvable points across the diameter of the hologram must be

$$N > 8 \times 2S_I$$
$$> 1120. \tag{10.16}$$

11

Applications in imaging

11.1 Particle-size analysis

Measurements on small, moving particles distributed through an appreciable volume are not possible with a conventional optical system because a microscope which can resolve particles of diameter d has a limited depth of field

$$\Delta z = d^2/2\lambda. \tag{11.1}$$

This problem can be overcome by using a hologram recorded with a pulsed laser to store a high-resolution, three-dimensional image of the whole field at any instant. The stationary image reconstructed by the hologram can then be examined in detail, at different levels, with a normal microscope [Thompson, Ward & Zinky, 1967].

Wherever the amount of light that is directly transmitted is large enough (>80 percent) to serve as a reference beam, it is possible to use in-line holography. This permits a very simple optical system, such as that shown in fig. 11.1, which is also economical of light.

Because of the very small diameter of the particles, the distance z of the recording plane from the particles can be made to satisfy the far-field

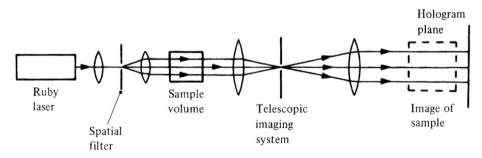

Fig. 11.1.   In-line holographic system for particle-size analysis.

94

condition, $(z \gg d^2/\lambda)$ (see Appendix C), quite easily. Such a hologram formed in the far field of the particles by the interference of the diffracted light and the directly transmitted light is known as a Fraunhofer hologram.

The permissible exposure time for recording a hologram of a moving particle field depends on the velocity of the particles. For size analysis, a useful criterion is that the particle should not move by more than a tenth of its diameter during the exposure. Suitable light sources are either a pulsed ruby laser or, where a higher repetition rate is necessary, a frequency-doubled Nd:YAG laser.

To produce an acceptable image of a particle, the hologram must record the central maximum and at least three side bands of its diffraction pattern. This would correspond to recording waves traveling at a maximum angle

$$\theta_{max} = 4\lambda/d \qquad (11.2)$$

to the directly transmitted wave, and, hence, to a maximum fringe frequency of $4/d$, which is independent of the values of λ and z.

It also follows from (11.2) that, for a given half-width of the hologram, x_{max}, the maximum depth of field over which the required resolution can be obtained is given by the relation

$$\Delta z_{max} = x_{max}\, d/4\lambda. \qquad (11.3)$$

To view the image, the hologram is illuminated with a collimated beam of light from a He–Ne laser. With normal processing, negative images are formed, but this is not a problem for most technical applications. Two images are formed at equal distances $\pm z$ from the hologram, one in front of it and the other behind it. However, with a Fraunhofer hologram, the light contributing to the conjugate image of each particle is spread over such a large area in the plane of the primary image that it produces only a weak, uniform background. As a result, the primary image can be viewed without significant interference from the conjugate image.

Holographic particle size analysis has found several applications [Trolinger, 1975; Vikram, 1992] including studies of fog droplets, dynamic aerosols and marine plankton. Another significant application has been in bubble-chamber photography. Double-exposure holography has been used to measure the velocity distribution of moving particles [Ewan, 1979].

11.2 Imaging through moving scatterers

Holography can be used to record an image of a stationary object masked by moving scatterers [Stetson, 1967]. Since the scattered light has its frequency shifted, it cannot interfere with the reference beam and merely adds a constant exposure to the hologram plate. Only the directly transmitted light contributes to the formation of the hologram.

11.3 Imaging through distorting media

Holography can be used to produce an undistorted image of an object which is located behind a distorting medium.

One way is to make the reference wave undergo the same distortion as the object wave [Goodman *et al.*, 1966]. This is possible if the distorting medium is very thin, or if the angular separation of the waves is very small.

If the distorting medium only modulates the phase of an incident wave, its amplitude transmittance can be written as $\exp[-i\phi(x, y)]$, and the complex amplitudes of the object and reference waves at the hologram are, respectively, $o(x, y)\exp[-i\phi(x, y)]$ and $r(x, y)\exp[-i\phi(x, y)]$. If we assume linear recording, the amplitude transmittance of the hologram is

$$\begin{aligned} t(x, y) &= t_0 + \beta T |r(x, y) \exp[-i\phi(x, y)] + o(x, y) \exp[-i\phi(x, y)]|^2 \\ &= t_0 + \beta T |r(x, y) + o(x, y)|^2, \end{aligned} \tag{11.4}$$

which is unaffected by the phase variations due to the distorting medium. An undistorted image is formed when the hologram is illuminated by the undistorted reference wave.

Alternatively, as shown in fig. 11.2(*a*), we can use a collimated reference beam to record a hologram of the aberrated object wave, whose complex amplitude we take to be

$$o(x, y) = |o(x, y)| \exp[-i\phi(x, y)]. \tag{11.5}$$

In the reconstruction step, the hologram is illuminated, as shown in fig. 11.2(*b*) by the conjugate of the original reference wave. The hologram then reconstructs the conjugate of the original object wave, whose complex amplitude in the hologram plane can be written, apart from a constant factor, as

$$o^*(x, y) = |o(x, y)| \exp[i\phi(x, y)]. \tag{11.6}$$

This wave has exactly the same phase errors as the original object wave, except that they are of the opposite sign. Accordingly, when this wave propagates back through the distorting medium, the phase errors introduced by it cancel out exactly, so that an undistorted real image of the object is formed in its original position [Kogelnik, 1965].

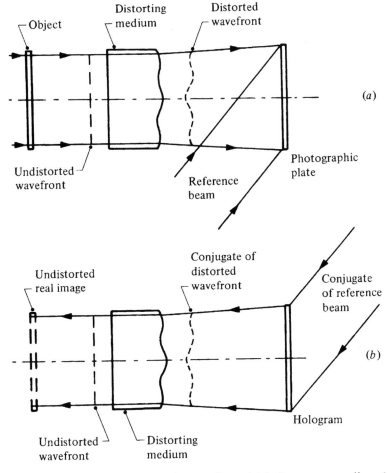

Fig. 11.2.   Imaging through a distorting medium: (*a*) hologram recording, (*b*) image reconstruction.

11.4 Time-gated imaging

A hologram records information on the object wave only when it is illuminated simultaneously by a coherent reference wave. If we use light with a limited coherence length, even with a cw source, only those parts of the object for which the difference in the lengths of the optical paths is less than the coherence length will be reconstructed in the image. It follows that a short coherence length produces results similar to a short light pulse [Abramson, 1978].

If a flat object surface and a hologram plate are both illuminated at an oblique angle by cw radiation with a short coherence length, the reconstructed image seen from any point on the hologram is crossed by a bright fringe connecting points at which the difference in the optical paths for the object and

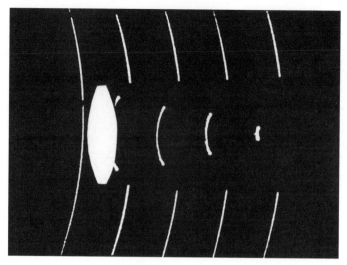

Fig. 11.3.   A time series (left to right) of images of a 3 mm thick (equivalent duration 10 ps) spherical wavefront focused by a converging lens [Abramson, 1983].

reference beams is close to zero. As the point of observation is moved along the hologram, the fringe moves to satisfy this condition, producing, as shown in fig. 11.3, a picture of the movement of the object wave [Abramson, 1983]. Light-in-flight recordings have also been made of a single pulse with a duration of 25 ps [Abramson & Spears, 1989].

Another application of time-gated imaging has been to obtain profiles of objects hidden within a scattering medium. This technique opens up the possibility of imaging through living tissues. A drawback is that only a small fraction of the light actually contributes to the formation of the hologram; the remainder forms an incoherent background. Methods for improving the contrast of the fringes have been reviewed by Chen *et al.* [1993].

11.5 Multiple imaging

There are many applications where it is necessary to produce an array of identical images. A hologram can be used to produce such an array with a single exposure.

11.5.1 Multiple imaging using Fourier holograms

An $n \times n$ array of identical images separated by intervals (x_0, y_0) can be produced by a hologram with an amplitude transmittance

$$H(\xi, \eta) = \sum^{n} \sum^{m} \exp[-\,i2\pi(nx_0\xi + my_0\eta)]. \tag{11.7}$$

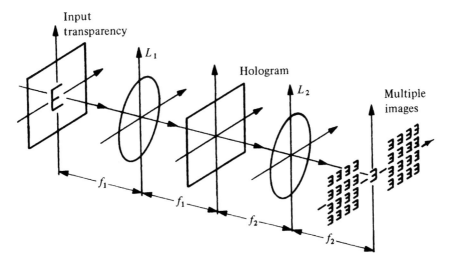

Fig. 11.4.   Multiple imaging by a Fourier hologram [Lu, 1968].

When illuminated with a plane wavefront, this hologram reconstructs a set of plane waves traveling in directions corresponding to the centers of the images in the array.

The hologram is placed in the back focal plane of the lens L_1 in the optical system shown in fig. 11.4 [Lu, 1968]. If a transparency with an amplitude transmittance $f(x, y)$ located in the front focal plane of L_1 and illuminated by a collimated beam is used as the input, its Fourier transform $F(\xi, \eta)$ is displayed in the back focal plane of L_1, so that the wavefront emerging from the hologram is

$$G(\xi, \eta) = F(\xi, \eta)\, H(\xi, \eta). \tag{11.8}$$

A second Fourier transform operation by the lens L_2 then produces a set of multiple images

$$g(x, y) = f(x, y) * \sum^{n}\sum^{m} \delta(x - nx_0, y - my_0)$$

$$= \sum^{n}\sum^{m} f(x - nx_0, y - my_0). \tag{11.9}$$

11.5.2 Multiple imaging using lensless Fourier holograms

This technique [Groh, 1968] uses a much simpler optical arrangement. In the first step a Fourier hologram of an array of point sources $P_1 \ldots P_n$ is recorded

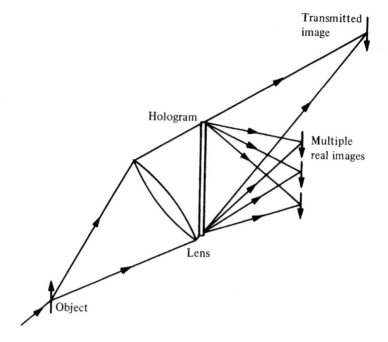

Fig. 11.5.   Multiple imaging by a lensless Fourier hologram [Groh, 1968].

with a point reference source. This hologram is then illuminated, as shown in fig. 11.5, with the conjugate to the original reference wave by means of a lens placed behind it.

When illuminated with a point source in this manner, the hologram produces real images of the array of object points $P_1 \ldots P_n$ in their original positions. However, if the point source is replaced by an illuminated transparency located in the same plane, an array of images of the transparency is formed, centered on the positions of the original point sources $P_1 \ldots P_n$.

References

Abramson, N. (1978). Light-in-flight recording by holography. *Optics Letters*, **3**, 121–3.

Abramson, N. H. (1983). Light-in-flight recording: high-speed holographic motion pictures of ultrafast phenomena. *Applied Optics*, **22**, 215–32.

Abramson, N. H. & Spears, K. G. (1989). Single pulse light-in-flight recording by holography. *Applied Optics*, **28**, 1834–41.

Chen, Y., Chen, H., Dilworth, D., Leith, E., Lopez, J., Shih, M., Sun, P. C. & Vossler, G. (1993). Evaluation of holographic methods for imaging through biological tissue. *Applied Optics*, **32**, 4330–6.

Ewan, B. C. R. (1979). Particle velocity distribution measurement by holography. *Applied Optics*, **18**, 3156–60.

Goodman, J. W., Huntley, W. H., Jackson, D. W. & Lehmann, M. (1966). Wavefront reconstruction imaging through random media. *Applied Physics Letters*, **8**, 311–13.

Groh, G. (1968). Multiple imaging by means of point holograms. *Applied Optics*, **7**, 1643–4.

Kogelnik, H. (1965). Holographic image projection through inhomogeneous media. *Bell System Technical Journal*, **44**, 2451–5.

Lu, S. (1968). Generating multiple images for integrated circuits by Fourier transform holograms. *Proceedings of the IEEE*, **56**, 116–17.

Stetson, K. A. (1967). Holographic fog penetration. *Journal of the Optical Society of America*, **57**, 1060–1.

Thompson, B. J., Ward, J. H. & Zinky, W. R. (1967). Application of hologram techniques for particle size analysis. *Applied Optics*, **6**, 519–26.

Trolinger, J. D. (1975). Particle field holography. *Optical Engineering*, **14**, 383–92.

Vikram, C. (1992). *Particle Field Holography*. Cambridge: Cambridge University Press.

Problems

11.1. Holographic imaging is to be used to study the size and spatial distribution of particles with diameters down to 10 μm, moving with velocities up to 1 m/s, in a field with a depth of 5 mm. Mechanical constraints require the hologram plate to be at a distance of 5 mm from the near side of the field. Determine (*a*) the minimum size of the hologram, (*b*) the minimum resolving power of the recording material, and (*c*) the maximum permissible exposure time.

The far side of the field is at a distance of 10 mm from the hologram. Accordingly, from (11.3) we require a hologram with a half-width

$$x_{max} = z_{max} \left(4\lambda/d\right)$$
$$= \frac{10 \times 10^{-3} \times 4 \times 0.694 \times 10^{-6}}{10 \times 10^{-6}} \text{ m}$$
$$= 2.78 \text{ mm.} \tag{11.10}$$

From (11.2), the maximum fringe frequency to be recorded is $4/0.01 = 400$ lines/mm. The resolving power of the film used must be significantly greater than this value.

During the exposure, a particle should not move by more than a tenth of its diameter. The maximum exposure time is, therefore, 1 μs.

12

Other applications

12.1 Holographic diffraction gratings

High quality diffraction gratings can be produced by recording an interference pattern in a layer of photoresist coated on an optically worked substrate [Schmahl & Rudolph, 1976; Hutley, 1982]. Holographic gratings, as they are commonly called, are free from periodic and random errors and exhibit very low levels of scattered light.

Figure 12.1 shows a typical optical system used to produce holographic gratings. Light from an Ar^+ laser ($\lambda = 458$ nm) is split into two beams of equal intensity which are focused by microscope objectives on pinholes. Each pinhole

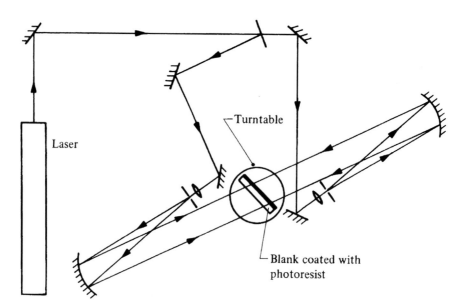

Fig. 12.1. Optical system used to produce blazed holographic gratings [courtesy I. G. Wilson, CSIRO Division of Chemical Physics, Melbourne, Australia].

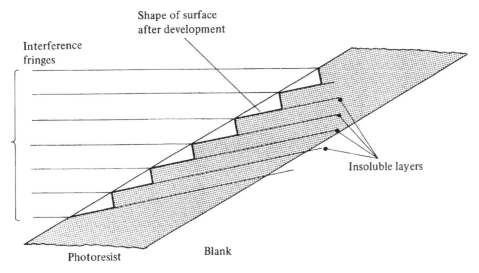

Fig. 12.2.   Production of blazed diffraction gratings in a photoresist [Hutley, 1982].

is placed at the focus of an off-axis parabolic mirror, so that two collimated beams are obtained upon reflection at the two mirrors.

As shown in fig. 12.2, these gratings can be blazed to obtain maximum diffraction efficiency for a specified wavelength by setting the photoresist layer obliquely to the fringe pattern, so as to generate a triangular groove profile [Hutley, 1976]. After processing, the surface of the photoresist is coated with an evaporated metal layer.

12.2 Holographic scanners

Holographic scanners are significantly cheaper than mirror scanners and are widely used in point of sale terminals and for high-speed nonimpact printing [Beiser, 1988]. A typical scanner consists, as shown in fig. 12.3, of a disc with a number of holograms recorded on it, using a point source as the object and a collimated reference beam. Rotating the disc causes the reconstructed image spot to scan the image plane.

With this simple arrangement, the scanning line is an arc of a circle with a radius

$$r = d \sin \theta \tan \theta, \qquad (12.1)$$

where d is the distance from the scanning facet to the center of the image plane, and θ is the angle between the normal to the hologram and the diffracted principal ray.

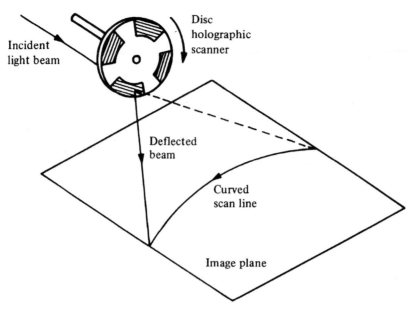

Fig. 12.3.   Disc holographic scanner [Kramer, 1981].

A straight line scan can be obtained by using an auxiliary reflector to make the principal diffracted ray normal to the scanned surface, as shown in fig. 12.4. This system is also self-compensating for wobble of the scanning disc when $\theta_i = \theta_d = 45°$.

12.3 Holographic filters

Volume reflection holograms recorded in dichromated gelatin can be used as narrow-band rejection filters (notch filters). To make such filters, a beam of laser light (wavelength λ), refracted at an angle θ within a layer of dichromated gelatin (refractive index n), is reflected off a mirror contacted to its back surface to produce interference fringes parallel to the surface with a spacing

$$\Lambda = \lambda / 2n \cos \theta. \tag{12.2}$$

The angle θ is chosen so that, after processing (see Section 6.2), the filter has its peak reflectance at the desired wavelength.

A major application of such notch filters has been for eye protection against laser radiation, while maintaining high visual transmittance [Magariños & Coleman, 1987; Tedesco, 1989]. Holographic notch filters are also used in Raman spectroscopy, to suppress the Rayleigh line while freely transmitting the Stokes region [Carraba *et al.*, 1990; Rich & Cook, 1991].

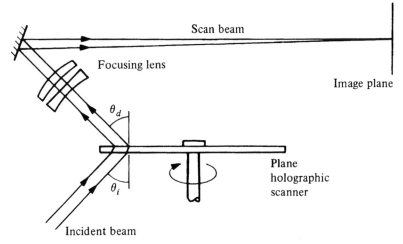

Fig. 12.4. Plane-grating holographic scanner producing a straight line scan [Kramer, 1981].

12.4 Holographic optical elements

A hologram can be used to transform an optical wavefront in the same manner as a lens. In addition, computer-generated holograms (see Chapter 10) can produce a wavefront having any arbitrary shape. As a result, holographic optical elements (HOEs) can perform unique functions and have been used in several specialized applications.

A major advantage of HOEs over conventional optical elements is that their function is independent of substrate geometry. In addition, since they can be produced on thin substrates, they are quite light, even for large apertures. Another advantage is that several holograms can be recorded in the same layer, so that spatially overlapping elements are possible. Finally, HOEs provide the possibility of correcting system aberrations in a single element, so that separate corrector elements are not required.

The recording material for a HOE must have high resolution, good stability, high diffraction efficiency and low scattering. Photoresists and dichromated gelatin are, at present, the most widely used materials. Photopolymers are an attractive alternative.

12.4.1 Head-up displays

One of the most successful applications of HOEs has been in head-up displays for high-performance aircraft, where, as shown in fig. 12.5, a HOE, used as a combiner, projects an image of the instruments at infinity, along the pilot's normal line of vision.

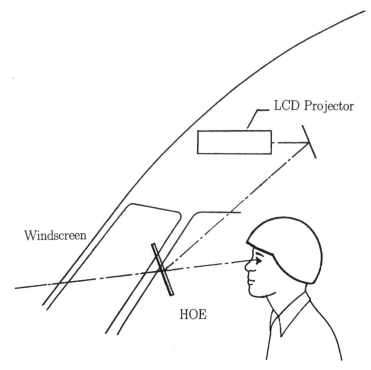

LCD Projector

Windscreen

HOE

Fig. 12.5.   Optical system used in an aircraft head-up display.

Holographic optical elements are lighter and can be fitted into the limited space available. In addition, a holographic combiner can be made with a high reflectance over a narrow band of wavelengths, and high transmittance at all other wavelengths [McCauley, Simpson & Murbach, 1973]. A design procedure for a curved combiner for a wide-field display has been described by Fisher [1989].

12.4.2 Beam shaping

Holographic optical elements are now used widely with laser diodes to correct the divergence and astigmatism of the beam [Hatakoshi & Goto, 1985]. Because of the difference in the recording and readout wavelengths, it is necessary to record the hologram with an aberrated wavefront [Amitai, Friesem & Weiss, 1990].

Two HOEs can also be used to generate a uniform circular or rectangular beam [Han, Ishii & Murata, 1983]. Another interesting application has been to generate beams with an amplitude profile described by a Bessel function [Vasara, Turunen & Friberg, 1989]. Such a beam has the property that its intensity profile does not change as it propagates, making it very useful for precision alignment. Yet another application has been in optical heads for compact-disc players [Lee, 1989]. Typically, the HOE produces three focused spots on the disc surface. The center spot is used to focus the beam and read out information, and the two outer spots provide a tracking error signal.

12.5 Interconnection networks

Integrated circuits in a computer are traditionally connected by metallic wires. Optical interconnections using holographic optical elements [Goodman *et al.* 1984] minimize propagation delays; in addition, they reduce space requirements, since several signals can propagate through the same network without mutual interference.

Holographic optical elements for interconnections must have high diffraction efficiency and SNR and should be capable of space-variant imaging, so that each source in an array can have a different interconnection pattern. New possibilities have been opened up by the development of laser-beam writing systems which can be used to fabricate highly efficient fanout elements [Prongué *et al.*, 1992].

Two-dimensional interconnection networks, such as perfect-shuffle networks, which require the spatial permutation of a $P \times Q$ array of light beams have potential applications in operations such as a matrix transpose or a fast Fourier transform. A space-variant optical interconnection system can be fabricated, as shown in fig. 12.6, with a pair of HOEs, each consisting of an array of holograms [Robertson *et al.*, 1991]. The input signals from an array of optical switches are redirected by the elements of H_1 to produce the required spatial permutation at the elements of H_2 which focuses them on an array of detectors.

However, such a system is not space-efficient and is sensitive to variations in the wavelength of the source. These problems are overcome in the compact implementation shown in fig. 12.7 [Song *et al.*, 1993], in which a single computer-generated HOE in a planar optical configuration is used to redirect light from an array of laser sources to an array of detectors.

12.6 Holographic memories

The maximum useful storage density with conventional techniques is set by the fact that dust or scratches can result in total loss of significant parts of the information. With holograms, surface damage only results in a drop in the overall signal-to-noise ratio, making much higher storage densities possible.

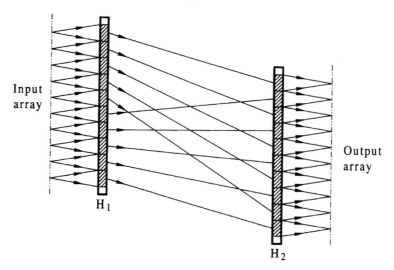

Fig. 12.6.   Space-variant perfect shuffle interconnection network using a pair of space-variant HOEs [Robertson *et al.*, 1991].

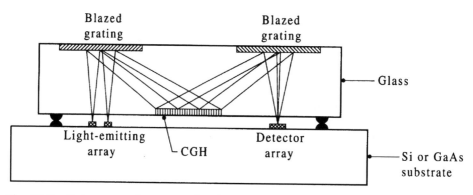

Fig. 12.7.   Interconnection network using a planar optical configuration [Song *et al.*, 1993].

This led, even quite early, to the development of page-organized holographic memories, such as the one shown schematically in fig. 12.8, in which two acousto-optic modulators were used to deflect a laser beam to address any one of an array of holograms. The information stored in the hologram was then read out by a detector array in the image plane.

Since there is not much scope for reducing the cycle time of holographic memories, subsequent work has been mainly directed towards increasing their capacity. One approach has been to use the fact that several holograms can be recorded in the same thick recording medium and read out separately [D'Auria *et al.*, 1974; Sincerbox, 1996].

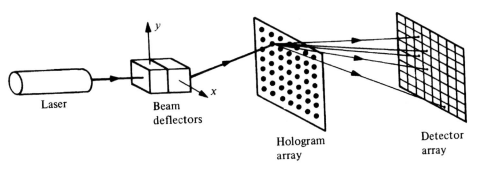

Fig. 12.8. Optical system for a read-only page-organized holographic information store [Kogelnik, 1972].

Another advantage of holographic information storage is the possibility of associative retrieval. Illumination of the memory plane with a search code produces an image in the detector plane for those holograms which contain a record of a logical match for the search code. This information can then be used to steer the reference beam to each of these holograms, in turn, to read out the data stored in it.

12.7 Holographic neural networks

Holographic neural networks are attractive because they offer large storage capacity as well as parallel access and processing capabilities during both the learning and reading phases [Psaltis, Brady & Wagner, 1988].

In a holographic neural network, neurons are represented by the pixels on a spatial light modulator. The brightness of a pixel corresponds to the activation level of the neuron. If a pair of pixels are illuminated with a coherent beam, a volume hologram can be formed in a suitable recording material. If, subsequently, one of the original two beams is used to address the hologram, the other beam is reconstructed with an efficiency that represents the weight between these neurons. With a photorefractive recording material, a process of learning can be implemented by increasing, or decreasing, the weights selectively.

References

Amitai, Y., Friesem, A. A. & Weiss, V. (1990). Designing holographic lenses with different recording and readout wavelengths. *Journal of the Optical Society of America. A*, **7**, 80–6.

Beiser, L. (1988). *Holographic Scanning*. New York: Wiley.

Carraba, M. M., Spencer, K. M., Rich, C. & Rault, D. (1990). The utilization of a holographic Bragg diffraction filter for Rayleigh line rejection in Raman spectroscopy. *Applied Spectroscopy*, **44**, 1558–61.

D'Auria, L., Huignard, J. P., Slezak, C. & Spitz, E. (1974). Experimental holographic read-write memory using 3-D storage. *Applied Optics*, **13**, 808–18.

Fisher, R. L. (1989). Design methods for a holographic head-up display curved combiner. *Optical Engineering*, **28**, 616–21.

Goodman, J. W., Leonberger, F. J., Kung, S.-Y. & Athale, R. A. (1984). Optical interconnections for VLSI systems. *Proceedings of the IEEE*, **72**, 850–66.

Han, C.-Y., Ishii, Y. & Murata, K. (1983). Reshaping collimated laser beams with Gaussian profile to uniform profiles. *Applied Optics*, **22**, 3644–7.

Hatakoshi, G. & Goto, K. (1985). Grating lenses for the semiconductor laser wavelength. *Applied Optics*, **24**, 4307–11.

Hutley, M. C. (1976). Interference (holographic) diffraction gratings. *Journal of Physics E: Scientific Instruments*, **9**, 513–20.

Hutley, M. C. (1982). *Diffraction Gratings*. London: Academic Press.

Kogelnik, H. (1972). Optics at Bell Laboratories – lasers in technology. *Applied Optics*, **11**, 2426–34.

Kramer, C. J. (1981). Holographic laser scanners for nonimpact printing. *Laser Focus*, **17**, No. 6, 70–82.

Lee, W. H. (1989). Holographic optical head for compact disk applications. *Optical Engineering*, **28**, 650–3.

Magariños, J. R. & Coleman, D. J. (1987). Holographic optical configuration for eye protection against lasers. *Applied Optics*, **26**, 2575–81.

McCauley, D. G., Simpson, C. E. & Murbach, W. J. (1973). Holographic optical element for visual display applications. *Applied Optics*, **12**, 232–42.

Prongué, D., Herzig, H. P., Dändliker, R. & Gale, M. T. (1992). Optimized kinoform structures for highly efficient fan-out elements. *Applied Optics*, **31**, 5706–11.

Psaltis, D., Brady, D. & Wagner, K. (1988). Adaptive optical networks using photorefractive crystals. *Applied Optics*, **27**, 1752–9.

Rich, C. & Cook, D. (1991). Lippmann volume holographic filters for Rayleigh line rejection in Raman spectroscopy. In *Practical Holography V*, Proceedings of the SPIE, vol. 1461, ed. S. A. Benton, pp. 2–7. Bellingham: SPIE.

Robertson, B., Restall, E. J., Taghizadeh, M. R. & Walker, A. C. (1991). Space-variant holographic optical elements in dichromated gelatin. *Applied Optics*, **30**, 2368–75.

Schmahl, G. & Rudolph, D. (1976). Holographic diffraction gratings. In *Progress in Optics*, vol. 14, ed. E. Wolf, pp. 196–244. Amsterdam: North-Holland.

Sincerbox, G. T., ed. (1996). *Selected Papers on Holographic Storage*, SPIE Milestone Series, vol. MS95. Bellingham: SPIE.

Song, S. H., Carey, C. D., Selviah, D. R., Midwinter, J. E. & Lee, E. H. (1993). Optical perfect shuffle interconnection using a computer-generated hologram. *Applied Optics*, **32**, 5022–5.

Tedesco, J. M. (1989). Holographic laser-protective filters and eyewear. *Optical Engineering*, **28**, 609–15.

Vasara, A., Turunen, J. & Friberg, A. T. (1989). Realization of general nondiffracting beams with computer-generated holograms. *Journal of the Optical Society of America. A*, **6**, 1748–54.

Problems

12.1. We need a holographic narrow-band notch filter to reject the light from a pulsed ruby laser ($\lambda = 694$ nm). This filter is to be produced in a dichromated gelatin layer using a collimated beam from an Ar^+ laser ($\lambda = 488$ nm). What is the angle at which the beam should be incident on the filter blank?

The filter is produced by reflecting the incident laser beam off a mirror contacted to the back surface of the filter blank. To reject a wavelength of 694 nm at normal incidence, the spacing of the fringe planes produced by the two interfering beams in the dichromated gelatin layer ($n = 1.48$) must be

$$\Lambda = \frac{694}{2 \times 1.48}$$
$$= 234.5 \text{ nm.} \tag{12.3}$$

From (12.2), the angle of incidence of the beam from the Ar^+ laser, within the dichromated gelatin layer, is given by the relation,

$$\cos \theta_n = \frac{488}{2 \times 1.48 \times 234.5}$$
$$= 0.703, \tag{12.4}$$

so that $\theta_n = 45.3°$.

Since θ_n is greater than the critical angle (42.5°), the filter blank cannot be illuminated directly with the laser beam. Instead, the gelatin layer must be contacted to a prism (in this case, a 45° prism could be used) with an index-matching fluid during recording. If the material of the prism has the same refractive index (1.48) as the gelatin layer, the angle of incidence of the laser beam on the other face of the prism should be

$$\theta = 1.48 \, (45.3° - 45°)$$
$$= 0.44°. \tag{12.5}$$

Because the incident beam is totally reflected at the back surface of the blank, there is no need to contact a mirror to it.

13

Holographic interferometry

Holography makes it possible to store a wavefront and reconstruct it at a later time. As a result, interferometric techniques can be used to compare two wavefronts which were originally separated in time or space, or even wavefronts of different wavelengths. In addition, since a hologram reconstructs the shape of an object with a rough surface faithfully, down to its smallest details, large scale changes in the shape of almost any object can be measured with interferometric precision [Brooks, Heflinger & Wuerker, 1965; Burch, 1965; Collier, Doherty & Pennington, 1965; Haines & Hildebrand, 1965; Stetson & Powell, 1965]. Holographic interferometry is now used extensively in nondestructive testing, aerodynamics, heat transfer and plasma diagnostics [Vest, 1979; Rastogi, 1994] as well as in studies of the behavior of anatomical structures and prostheses under stress [Greguss, 1975; von Bally, 1979; Podbielska, 1991, 1992].

13.1 Real-time interferometry

Equations (1.6)–(1.9) show that if a hologram is replaced in its original position in the same optical system used to record it, and illuminated with the original reference wave, it reconstructs the original object wave. If, then, the shape of the object changes slightly, the directly transmitted object wave will interfere with the reconstructed object wave to produce, as shown in fig. 13.1, a fringe pattern that maps the changes in the shape of the object.

If the changes in the shape of the object are small, only the phase of the object wave is modified, and the complex amplitude of the wave from the deformed object can be written as

$$o'(x, y) = o(x, y) \exp[-i\Delta\phi(x, y)], \qquad (13.1)$$

where $o(x, y)$ is the complex amplitude of the original object wave, and $\Delta\phi(x, y)$ is the phase change at a point (x, y) arising from the deformation. It then

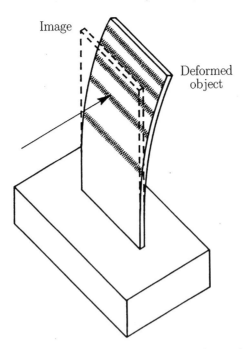

Fig. 13.1.   Generation of interference fringes by deformation of an object.

follows from (1.9) that the complex amplitude of the directly transmitted object wave, which only involves the first term on the right-hand side of (1.8), is

$$u_1(x, y) = (t_0 + \beta T r^2)\, o'(x, y). \tag{13.2}$$

Similarly, from (1.9), the complex amplitude of the reconstructed object wave is

$$u_2(x, y) = \beta T r^2 o(x, y). \tag{13.3}$$

The intensity in the interference pattern produced by these two waves is, therefore,

$$\begin{aligned}
I(x, y) &= |u_1(x, y) + u_2(x, y)|^2 \\
&= |o(x, y)|^2 [\beta^2 T^2 r^4 + (t_0 + \beta T r^2)^2 \\
&\quad + 2\beta T r^2 (t_0 + \beta T r^2) \cos \Delta\phi(x, y)].
\end{aligned} \tag{13.4}$$

Since β is negative, dark fringes are seen when $\Delta\phi(x, y) = 2m\pi$, where m is an integer.

Interference fringes obtained by this technique can be used to study changes in the shape of the object in real time. The problem of replacing the hologram in its original position can be eliminated by *in situ* processing of the hologram plate in a liquid gate with a monobath [Hariharan & Ramprasad, 1973], as

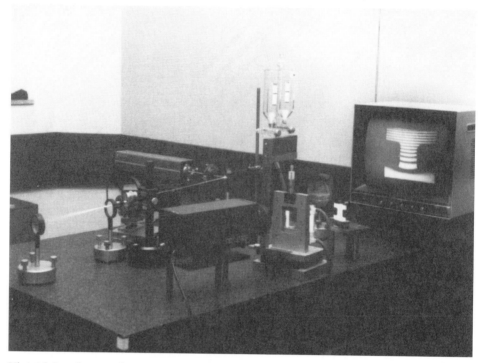

Fig. 13.2. System for real-time holographic interferometry. The hologram is processed *in situ* and the interference fringes are viewed through a closed-circuit television
system.

shown in fig. 13.2, or by using a photothermoplastic or a photorefractive
crystal as the recording material.

A problem in real-time holographic interferometry is that the light scattered
by the object is largely depolarized. The resulting drop in the visibility of the
fringes can be avoided by using a polarizer when viewing the fringes.

13.2 Double-exposure interferometry

It is also possible to record two holograms on the same photographic plate; one
of the object in its original state, and the other of the deformed object. The
resultant complex amplitude, due to the superposition of the two reconstructed images is then, apart from a constant of proportionality,

$$u(x, y) = o(x, y) + o'(x, y)$$
$$= o(x, y)\{1 + \exp[-i\Delta\phi(x, y)]\}, \tag{13.5}$$

and the intensity in the image is

$$I(x, y) = |o(x, y)|^2[1 + \cos\Delta\phi(x, y)]. \tag{13.6}$$

In this case, bright fringes are seen when $\Delta\phi(x, y) = 2m\pi$.

Double-exposure holographic interferometry has the advantage that repositioning of the hologram is not critical, since the two interfering waves are always reconstructed in exact register. In addition, the visibility of the fringes is good, since the two waves have the same polarization and the same amplitude.

A common problem in double-exposure holographic interferometry with a cw laser is unwanted movements of the object between the two exposures. Some types of object motion can be eliminated by reflecting the reference beam from a mirror attached to the object [Mottier, 1969]. Alternatively, the hologram plate can be attached to the object, and a doubly exposed reflection hologram can be recorded with the object illuminated through the hologram plate [Boone, 1975].

A disadvantage of double-exposure holographic interferometry is that information on intermediate states of the object is not available. In addition, control of the fringes, to compensate for rigid body motion and avoid ambiguities in interpretation, is not possible.

The first problem can be overcome by making a series of exposures at successive stages of loading, using a set of masks with apertures that overlap in a predetermined order [Hariharan & Hegedus, 1973]. The reconstructed images then yield interference patterns corresponding to any two stages of loading. An elegant solution to both these problems is the sandwich hologram

13.3 Sandwich holograms

In this technique [Abramson, 1974, 1975; Abramson & Bjelkhagen, 1979], as shown in fig. 13.3, pairs of holographic plates (without any antihalation backing) are exposed in the same plate holder, with their emulsion-coated surfaces facing the object. B_1, F_1 are exposed with the unstressed object, while B_2, F_2 and B_3, F_3, . . ., are exposed at successive stages of loading of the object. After the plates are processed, if B_1 is combined with F_2, F_3, . . ., in the original plate holder and illuminated with the original reference beam, it is possible to study the total deformation at any stage, while combinations such as B_2F_3, B_3F_4, . . . show the incremental deformations.

In addition, as shown in fig. 13.4, ambiguities can be resolved by tilting the sandwich; this results in a change in the interference pattern exactly equivalent to tilting the object between the two exposures.

13.4 Industrial measurements

Holographic interferometry normally requires an extremely stable environment. However, various techniques have been developed which make it possible to use holographic interferometry in an industrial environment.

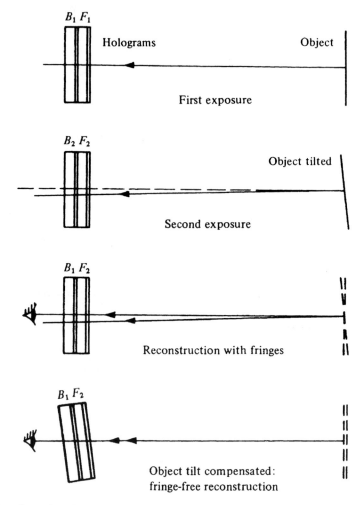

Fig. 13.3.   Steps involved in sandwich hologram interferometry [Abramson, 1975].

In some situations, holographic interferometry can be carried out with a cw source by reflecting the reference wave from a mirror attached to the object [Mottier, 1969].

However, the most common method is to use a double-pulsed laser. Double-exposure holographic interferometry can then be used to study transient phenomena, such as deformations due to impact loading [Gates, Hall & Ross, 1972; Armstrong & Forman, 1977]. An electrical, optical or acoustic signal is used to trigger the first pulse just before the instant of impact, with the second pulse following after a predetermined delay. Sandwich holography can be used to eliminate unwanted rigid body displacements [Bjelkhagen, 1977; Abramson & Bjelkhagen, 1978].

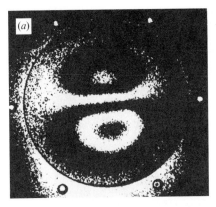 

Fig. 13.4. Interference fringes obtained, using a sandwich hologram, with a thin metal sheet clamped at its edges and subjected to a bending moment about a horizontal diameter: (*a*) with the sandwich hologram in its original position and (*b*) after tilting the sandwich hologram about a vertical axis [Hariharan & Hegedus, 1976].

Objects rotating at extremely high speeds (see fig. 13.5), can also be studied with the aid of an optical derotator [Stetson, 1978], consisting of an inverting prism that rotates at half the speed of the object.

13.5 Refractive index fields

Holographic interferometry has practical advantages even in applications such as flow visualization and heat transfer studies, where conventional interferometry has been used for many years. In the first instance, mirrors and windows of relatively low optical quality can be used, since the phase errors due to the optics contribute equally to both interfering wavefronts and, therefore, cancel out. However, the most significant advantage is that a diffusing screen can be used to obtain an interference pattern that is localized near the phase object and can be viewed over a range of angles. This makes it possible to study three-dimensional refractive index distributions [Sweeney & Vest, 1973].

Holographic interferometry has also been found useful in plasma diagnostics. Since a plasma is highly dispersive, measurements of the refractive index distribution at two wavelengths can be used to determine the electron density directly. One way to do this is to record two holograms simultaneously, on the same plate, with light from a ruby laser that has passed through a frequency doubler to produce two collinear beams with wavelengths of 694 nm and 347 nm. It is then possible to make the second-order image reconstructed by one hologram interfere with the first-order image reconstructed by the other. The interference fringes formed are contours of constant dispersion and, hence, of constant electron density [Ostrovskaya & Ostrovskii, 1971].

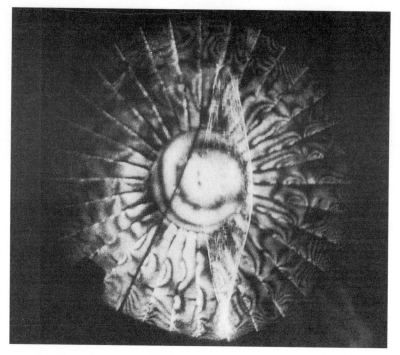

Fig. 13.5.   Holographic interferogram of a turbine fan rotating at 4460 rpm recorded with an image derotator and a double-pulsed ruby laser [courtesy K. A. Stetson, United Technology Research Center, East Hartford, USA].

13.6 Surface displacements

With an object having a rough surface, the phase varies in a random manner across the object wavefront. As a result, only waves from corresponding points on the object wavefront and the reconstructed wavefront contribute effectively to the interference pattern, and the intensity at any point in it is determined by the phase difference between the waves from these two points.

To evaluate this phase difference, we consider a point P on the surface which, as shown in fig. 13.6, has undergone a vector displacement **L** to P'.

If the displacement of P is small compared to the distances to the source S and the point of observation O, the phase difference introduced is

$$\Delta\phi = \mathbf{L} \cdot (\mathbf{k}_1 - \mathbf{k}_2)$$
$$= \mathbf{L} \cdot \mathbf{K}, \tag{13.7}$$

where $\mathbf{k}_1$ and $\mathbf{k}_2$ are the propagation vectors of the incident and scattered beams, and $\mathbf{K} = \mathbf{k}_1 - \mathbf{k}_2$ is known as the sensitivity vector [Aleksandrov & Bonch-Bruevich, 1967; Ennos, 1968; Sollid, 1969].

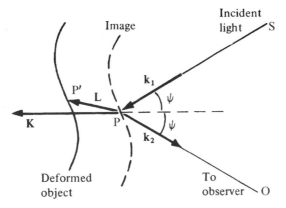

Fig. 13.6.   Evaluation of the phase difference produced by a local displacement of the object.

13.7 The holodiagram

The holodiagram is a useful aid to interpretation of the interference fringes [Abramson, 1969]. For the basic hologram recording system shown in fig. 13.7, the holodiagram consists, as shown in fig. 13.8, of a set of ellipses whose foci, O and O′, correspond, respectively, to the beam splitter and the viewing point on the photographic plate.

For any object point P, the ellipse on which it lies is the locus of P for which the distance OPO′ has a constant value. A displacement of P from one ellipse to the next corresponds to a change in this distance of one wavelength and a shift of one fringe in the interference pattern. The required displacement of P is obviously a minimum when its motion is along the normal to the ellipse, which corresponds to the sensitivity vector **K**.

The circles drawn through O and O′ are curves of constant **K**. They correspond to the specified values of the parameter $q = 1/\cos \psi$, where the angle OPO′ $= 2\psi$. These curves can be used to optimize a hologram recording system for measurements of a particular type of surface displacement.

13.8 Fringe localization

With an object having a rough surface, the visibility of the interference fringes is a maximum for a particular position of the plane of observation, known as the plane of localization.

As mentioned earlier, because of the random phase variations across the object wavefront, only waves from corresponding points on the two interfering wavefronts contribute effectively to the interference fringes. For a given viewing

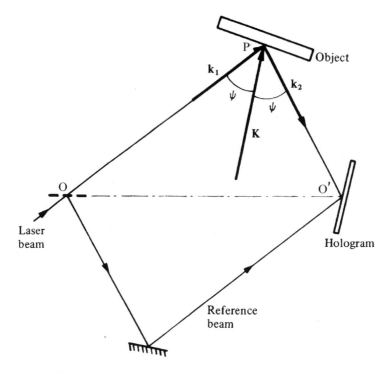

Fig. 13.7.   Schematic of a basic hologram recording system.

direction, the phase difference $\Delta\phi$ between the waves from two such points, P and P' (see fig. 13.6), is defined by (13.7). In general, this phase difference will vary over the range of viewing directions defined by the aperture of the viewing lens, resulting in a loss of contrast of the fringes. However, it is possible to find a plane in which the value of $\Delta\phi$ is very nearly constant over this range of viewing directions; this is the plane of localization of the fringes [Walles, 1969].

The position of the plane of localization depends on the type of displacement. Two cases are of particular interest. One is pure translation of the object, which produces fringes localized at infinity; the other is rotation of the object about an axis contained in its surface, which results in fringes localized at the surface [Molin & Stetson, 1970a,b].

13.9 Strain analysis

Inspection of the fringe pattern is quite useful to detect localized defects or areas of stress concentration. However, quantitative stress analysis requires measurements of the local strains.

If, at any point on the stressed object, L_x, L_y, and L_z denote the x, y, and z

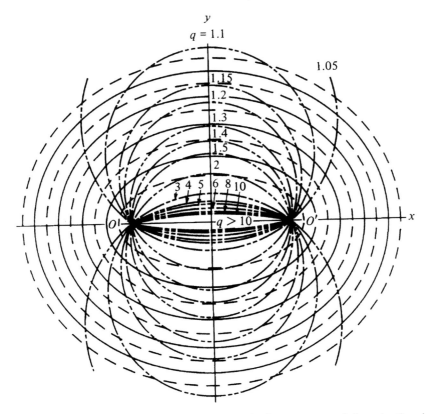

Fig. 13.8. The holodiagram. The ellipses are loci of constant path length; the circles are loci of constant **K** [Abramson, 1969].

components, respectively, of the surface displacement, the three components of the normal strain at this point are defined by the relations

$$\epsilon_x = \partial L_x / \partial x, \tag{13.8}$$
$$\epsilon_y = \partial L_y / \partial y, \tag{13.9}$$
$$\epsilon_z = \partial L_z / \partial z, \tag{13.10}$$

and the three shear strains are

$$\gamma_{xy} = (\partial L_x / \partial y) + (\partial L_y / \partial x), \tag{13.11}$$
$$\gamma_{yz} = (\partial L_y / \partial z) + (\partial L_z / \partial y), \tag{13.12}$$
$$\gamma_{zx} = (\partial L_z / \partial x) + (\partial L_x / \partial z). \tag{13.13}$$

Several methods of analysis have been proposed to evaluate the surface displacements and, hence, the strains [Briers, 1976].

Early workers tended to favor methods using observations of fringe localization, since, in some cases, they permit direct measurements of the strain [Dubas

& Schumann, 1975]. However, their accuracy is limited, and interpretation of the fringes is sometimes difficult.

The fringe vector method [Stetson, 1974, 1979] uses the fact that any combination of homogeneous deformation and rotation of an object yields fringes that appear to be produced by the intersection of the object surface with a number of equally spaced surfaces which are contours of constant phase difference. The fringe vector runs perpendicular to these surfaces, and its magnitude is inversely proportional to their separation.

To apply this method, the fringe vectors corresponding to three different directions of viewing are determined. The gradients of the displacements along the x, y, and z axes can then be evaluated directly from the resolved components of the fringe vectors and the sensitivity vectors along these axes [Pryputniewicz, 1978].

Another way to calculate the strains is to evaluate the surface displacements, with reference to some point in the field which is assumed to be stationary, and to differentiate these values. While it is possible to use three observations of the fringe order made from three directions, a better method is to use a single direction of observation and three different directions of object illumination [Hung *et al.*, 1973]. The measured phase differences $\Delta\phi_1$, $\Delta\phi_2$, and $\Delta\phi_3$ are then linked to L_x, L_y, and L_z, the three orthogonal components of the displacement by the matrix relation

$$
\begin{bmatrix} K_{1x} & K_{1y} & K_{1z} \\ K_{2x} & K_{2y} & K_{2z} \\ K_{3x} & K_{3y} & K_{3z} \end{bmatrix}
\begin{bmatrix} L_x \\ L_y \\ L_z \end{bmatrix} =
\begin{bmatrix} \Delta\phi_1 \\ \Delta\phi_2 \\ \Delta\phi_3 \end{bmatrix}, \tag{13.14}
$$

where $\mathbf{K}_1$, $\mathbf{K}_2$ and $\mathbf{K}_3$ are the values of the sensitivity vector for the three directions of illumination.

Data reduction can be simplified by illuminating the object from four different directions making equal angles with the viewing direction, two in the vertical plane and two in the horizontal plane [Goldberg, 1975].

References

Abramson, N. (1969). The holo-diagram: a practical device for making and evaluating holograms. *Applied Optics*, **8**, 1235–40.

Abramson, N. (1974). Sandwich hologram interferometry: a new dimension in holographic comparison. *Applied Optics*, **13**, 2019–25.

Abramson, N. (1975). Sandwich hologram interferometry. 2. Some practical calculations. *Applied Optics*, **14**, 981–4.

Abramson, N. & Bjelkhagen, H. (1978). Pulsed sandwich holography. 2. Practical application. *Applied Optics*, **17**, 187–9.

Abramson, N. & Bjelkhagen, H. (1979). Sandwich hologram interferometry. 5. Measurement of in-plane displacement and compensation for rigid body motion. *Applied Optics*, **18**, 2870–80.

Aleksandrov, E. G. & Bonch-Bruevich, A. M. (1967). Investigation of surface strains by the hologram technique. *Soviet Physics: Technical Physics*, **12**, 258–65.

Armstrong, W. T. & Forman, P. R. (1977). Double-pulsed time differential holographic interferometry. *Applied Optics*, **16**, 229–32.

Bjelkhagen, H. (1977). Pulsed sandwich holography. *Applied Optics*, **16**, 1727–31.

Boone, P. M. (1975). Use of reflection holograms in holographic interferometry and speckle correlation for measurement of surface displacement. *Optica Acta*, **22**, 579–89.

Briers, J. D. (1976). The interpretation of holographic interferograms. *Optics & Quantum Electronics*, **8**, 469–501.

Brooks, R. E., Heflinger, L. O. & Wuerker, R. F. (1965). Interferometry with a holographically reconstructed comparison beam. *Applied Physics Letters*, **7**, 248–9.

Burch, J. M. (1965). The application of lasers in production engineering. *The Production Engineer*, **44**, 431–42.

Collier, R. J., Doherty, E. T. & Pennington, K. S. (1965). Application of moire techniques to holography. *Applied Physics Letters*, **7**, 223–5.

Dubas, M. & Schumann, W. (1975). On direct measurement of strain and rotation in holographic interferometry using the line of complete localization. *Optica Acta*, **22**, 807–19.

Ennos, A. E. (1968). Measurement of in-plane surface strain by hologram interferometry. *Journal of Physics E: Scientific Instruments*, **1**, 731–4.

Gates, J. W. C., Hall, R. G. N. & Ross, I. N. (1972). Holographic interferometry of impact-loaded objects using a double-pulse laser. *Optics & Laser Technology*, **4**, 72–5.

Goldberg, J. L. (1975). A holographic interferometer for the measurement of the vector displacement of a slowly deforming rough surface. *Japanese Journal of Applied Physics*, **14**(Supplement 14-1), 253–8.

Greguss, P. (1975). *Holography in Medicine*. London: IPC Press.

Haines, K. A. & Hildebrand, B. P. (1965). Contour generation by wavefront reconstruction. *Physics Letters*, **19**, 10–11.

Hariharan, P. & Hegedus, Z. S. (1973). Simple multiplexing technique for double-exposure hologram interferometry. *Optics Communications*, **9**, 152–5.

Hariharan, P. & Hegedus, Z. S. (1976). Two-hologram interferometry: a simplified sandwich technique. *Applied Optics*, **15**, 848–9.

Hariharan, P. & Ramprasad, B. S. (1973). Rapid *in situ* processing for real-time holographic interferometry. *Journal of Physics E: Scientific Instruments*, **6**, 699–701.

Hung, Y. Y., Hu, C. P., Henley, D. R., & Taylor, C. E. (1973). Two improved methods of surface displacement measurements by holographic interferometry. *Optics Communications*, **8**, 48–51.

Molin, N. E. & Stetson, K. A. (1970a). Measurement of fringe loci and localization in hologram interferometry for pivot motion, in-plane rotation and in-plane translation. Part I. *Optik*, **31**, 157–77.

Molin, N. E. & Stetson, K. A. (1970b). Measurement of fringe loci and localization in hologram interferometry for pivot motion, in-plane rotation and in-plane translation. Part II. *Optik*, **31**, 281–91.

Mottier, F. M. (1969). Holography of randomly moving objects. *Applied Physics Letters*, **15**, 44–5.

Ostrovskaya, G. V. & Ostrovskii, Yu. I. (1971). Two-wavelength hologram method for studying the dispersion properties of phase objects. *Soviet Physics – Technical Physics*, **15**, 1890–2.

Podbielska, H., ed. (1991). *Holography, Interferometry & Optical Pattern Recognition in Biomedicine*, Proceedings of the SPIE, vol. 1429, Bellingham: SPIE.

Podbielska, H., ed. (1992). *Holography, Interferometry & Optical Pattern Recognition in Biomedicine II*, Proceedings of the SPIE, vol. 1647, Bellingham: SPIE.

Pryputniewicz, R. J. (1978). Holographic strain analysis: an experimental implementation of the fringe vector theory. *Applied Optics*, **17**, 3613–18.

Rastogi, P. K., ed. (1994). *Holographic Interferometry*. Berlin: Springer-Verlag.

Sollid, J. E. (1969). Holographic interferometry applied to measurements of small static displacements of diffusely reflecting surfaces. *Applied Optics*, **8**, 1587–95.

Stetson, K. A. (1974). Fringe interpretation for holographic interferometry of rigid body motions and homogeneous deformations. *Journal of the Optical Society of America*, **64**, 1–10.

Stetson, K. A. (1978). The use of an image derotator in hologram interferometry and speckle photography of rotating objects. *Experimental Mechanics*, **18**, 67–73.

Stetson, K. A. (1979). Use of projection matrices in hologram interferometry. *Journal of the Optical Society of America*, **69**, 1705–10.

Stetson, K. A. & Powell, R. L. (1965). Interferometric hologram evaluation and real-time vibration analysis of diffuse objects. *Journal of the Optical Society of America*, **55**, 1694–5.

Sweeney, D. W. & Vest, C. M. (1973). Reconstruction of three-dimensional refractive index fields from multidirectional interferometric data. *Applied Optics*, **12**, 2649–64.

Vest, C. M. (1979). *Holographic Interferometry*. New York: John Wiley.

von Bally, G., ed. (1979). *Holography in Medicine and Biology*. Berlin: Springer-Verlag.

Walles, S. (1969). Visibility and localization of fringes in holographic interferometry of diffusely reflecting surfaces. *Arkiv för Fysik*, **40**, 299–403.

Problems

13.1. A hologram is recorded of a flat circular diaphragm clamped by its edge over an opening in a pressure vessel and illuminated at 45° with a beam from a He–Ne laser ($\lambda = 633$ nm). The hologram is processed *in situ* and the diaphragm is viewed, along the normal to its surface, through the hologram. When the pressure in the vessel is increased slightly, three dark circular fringes are seen surrounding a dark spot at the center of the reconstructed image of the diaphragm. What is the deflection of the center of the diaphragm?

Since the edge of the diaphragm is fixed, and we have four fringes from the edge to the center, the phase difference at the center of the interference pattern produced by the change in pressure is

$$\Delta\phi = 4 \times 2\pi = 25.13 \text{ radians.} \tag{13.15}$$

We also know that the displacement of the center of the diaphragm must be along the normal to its surface. Accordingly, it follows from fig. 13.6 and (13.7) that the magnitude of the sensitivity vector is

$$\mathbf{K} = (2\pi/\lambda)(1 + \cos 45°)$$

$$= \frac{2\pi(1 + 0.7071)}{633 \times 10^{-9}}$$

$$= 16.94 \times 10^6 \ \mathrm{m}^{-1}. \tag{13.16}$$

The displacement of the center of the diaphragm is, therefore,

$$L(0,0) = \Delta\phi/\mathbf{K}$$

$$= \frac{25.13}{16.94 \times 10^6} \ \mathrm{m}$$

$$= 1.48 \ \mu\mathrm{m}. \tag{13.17}$$

14

Advanced techniques

14.1 Moiré interferometry

In this technique, images of a master object and a similar test object are superimposed by means of a suitable optical system, and a hologram is recorded of the resultant wave field. This hologram is processed and replaced in its original position in the recording system.

If both the objects are stressed, the intensity at any point in the image is

$$I = I_0 \left[1 - \cos\left(\frac{\phi_1 - \phi_2}{2}\right) \cos\left(\frac{\phi_1 + \phi_2}{2}\right) \right] \qquad (14.1)$$

where $\phi_1 = \mathbf{K} \cdot \mathbf{L}_1$, $\phi_2 = \mathbf{K} \cdot \mathbf{L}_2$, $\mathbf{K}$ is the sensitivity vector, and $\mathbf{L}_1$ and $\mathbf{L}_2$ are the vector displacements of the corresponding points on the two objects.

An observer sees moiré fringes generated by the term $\cos\left[(\phi_1 - \phi_2)/2\right]$ which contour the differences in the displacements of the two objects [Der, Holloway & Fourney, 1973].

14.2 Vibrating surfaces

Holographic interferometry can be used to map the amplitude of vibration of an object with a diffusely reflecting surface.

14.2.1 Stroboscopic interferometry

In this technique [Archbold & Ennos, 1968], a hologram of the vibrating object is recorded using a sequence of light pulses that are triggered at times Δt_1 and Δt_2 after the start of each vibration cycle. If the displacement of a point (x, y) on the object at time t is given by the relation

$$\mathbf{L}(x, y, t) = \mathbf{L}(x, y) \sin \omega t, \qquad (14.2)$$

the intensity in the reconstructed image is

$$I(x, y) = I_0(x, y)\{1 + \cos [\mathbf{K} \cdot \mathbf{L}(x, y) (\sin \omega \Delta t_1 - \sin \omega \Delta t_2)]\}, \quad (14.3)$$

where $\mathbf{K}$ is the sensitivity vector, and $I_0(x, y)$ is the intensity in the image when the object is at rest.

The hologram is equivalent to a double-exposure hologram recorded with the object in these two states of deformation.

14.2.2 Time-average interferometry

In this technique, a hologram is recorded of the vibrating surface with an exposure time that is long compared to the period of vibration [Powell & Stetson, 1965]. The amount by which the phase of the light from any point on the object is shifted is then a function of time and can be written as

$$\Delta\phi(x, y, t) = \mathbf{K} \cdot \mathbf{L}(x, y) \sin \omega t. \quad (14.4)$$

The complex amplitude of the scattered light wave from the vibrating object is, therefore,

$$o(x, y, t) = |o(x, y)| \exp\{-i[\phi(x, y) + \mathbf{K} \cdot \mathbf{L}(x, y) \sin \omega t]\}. \quad (14.5)$$

If the holographic recording process is linear, the complex amplitude $u(x, y)$ of the wave reconstructed by the hologram will be proportional to the time-average of $o(x, y, t)$ over the exposure interval T, so that we can write

$$\begin{aligned}
u(x, y) &= \frac{1}{T} \int_0^T |o(x, y)| \exp\{-i[\phi(x, y) + \mathbf{K} \cdot \mathbf{L}(x, y) \sin \omega t]\}\mathrm{d}t, \\
&= |o(x, y)| \exp[-i\phi(x, y)] \frac{1}{T} \int_0^T \exp[-i\mathbf{K} \cdot \mathbf{L}(x, y) \sin \omega t]\}\mathrm{d}t, \\
&= o(x, y)M_T(x, y), \quad (14.6)
\end{aligned}$$

where $M_T(x, y)$ is known as the characteristic function. If the exposure time is long compared to the period of the vibration $(T \gg 2\pi/\omega)$, we have

$$\begin{aligned}
M_T(x, y) &= T \overset{\lim}{\to} \infty \frac{1}{T} \int_0^T \exp[-i\mathbf{K} \cdot \mathbf{L}(x, y) \sin \omega t]\mathrm{d}t, \\
&= J_0[\mathbf{K} \cdot \mathbf{L}(x, y)], \quad (14.7)
\end{aligned}$$

where J_0 is the zero-order Bessel function of the first kind. The intensity in the reconstructed image is then

$$\begin{aligned}
I(x, y) &= |o(x, y) M_T(x, y)|^2 \\
&= I_0(x, y) J_0^2 [\mathbf{K} \cdot \mathbf{L}(x, y)]. \quad (14.8)
\end{aligned}$$

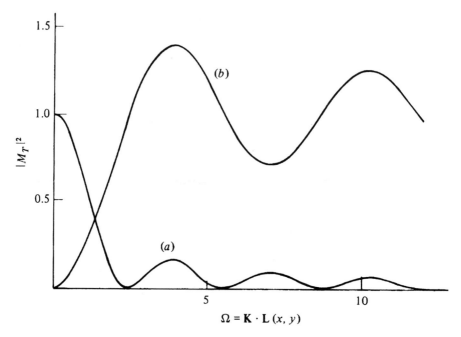

Fig. 14.1.   Characteristic functions for a vibrating object: (*a*) time-average fringes and
(*b*) real-time fringes.

The function $|M_T|^2$ is plotted against the parameter $\Omega = \mathbf{K} \cdot \mathbf{L}$ in fig. 14.1(*a*).
If the vibration amplitude varies over the object, (14.8) gives rise to fringes
(contours of equal vibration amplitude) covering the reconstructed image. The
dark fringes, at which the intensity drops to zero, correspond to the zeros of
the function $J_0^2(\Omega)$, and the bright fringes to its maxima. The first maximum,
which corresponds to the nodes, is the brightest, while the intensities of successive maxima, occurring at larger vibration amplitudes, fall off progressively.

A series of time-averaged interferograms obtained with an acoustic guitar,
showing the resonant modes of its sound board, is presented in fig. 14.2.

Time-averaged fringes can also be observed using real-time holographic interferometry [Stetson & Powell, 1965]. In this case, the characteristic function is

$$|M_T|^2 = 1 - J_0[\mathbf{K} \cdot \mathbf{L}(x, y)]. \tag{14.9}$$

The function defined by (14.9) is shown in fig. 14.1(*b*), while fig. 14.3 shows
typical fringe patterns obtained with the same object using the time-average
and real-time techniques [Biedermann & Molin, 1970]. As can be seen, only
half the number of fringes are seen with the real-time technique, and their contrast is much lower.

Real-time observations are most useful when searching for resonant modes,

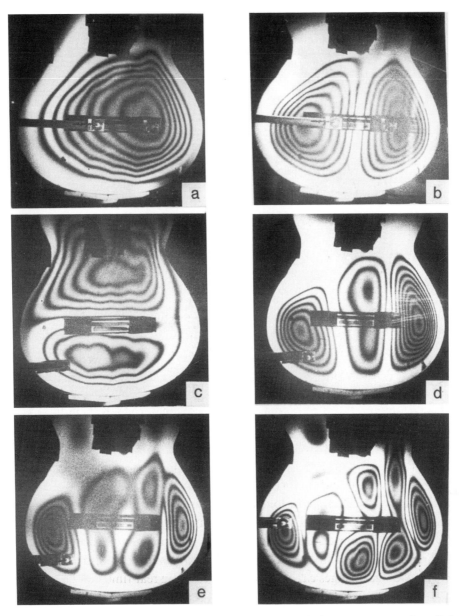

Fig. 14.2. Time-average holographic interferograms of the resonant modes of the soundboard of a guitar at frequencies of (*a*) 195, (*b*) 292, (*c*) 385, (*d*) 537, (*e*) 709 and (*f*) 905 Hz.

(a)

(b)

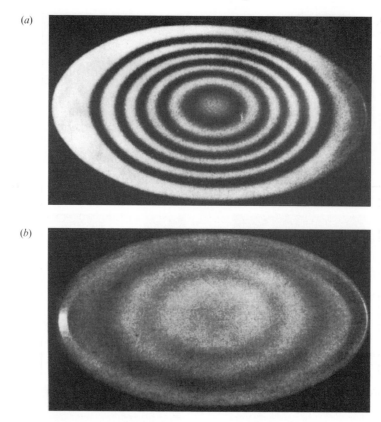

Fig. 14.3.   Fringe patterns obtained with the same vibrating object (a tin can) using (a) the time-average technique, and (b) real-time interference [Biedermann & Molin, 1970].

after which a time-average hologram can be made for detailed measurements of the vibration amplitude. Typical applications of time-average holographic interferometry have been in studies of musical instruments [Ågren & Stetson, 1972], loudspeakers [Chomat & Miler, 1973], turbine blades, and aircraft structures [Bjelkhagen, 1974].

14.3 Contouring

Holographic interferometry can produce an image of a three-dimensional object overlaid with contours of constant elevation.

14.3.1 *Two-wavelength contouring*

In this technique [Haines & Hildebrand, 1965; Zelenka & Varner, 1968], a telecentric system is used, as shown in fig. 14.4, to image the object on the

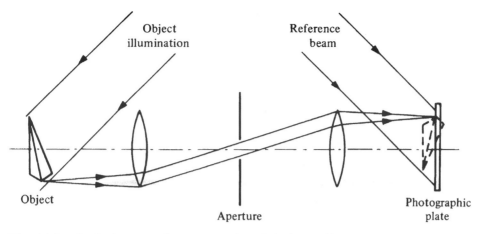

Fig. 14.4. Optical system for two-wavelength holographic contouring [Zelenka & Varner, 1968].

hologram. A collimated beam is used to illuminate the object, and another collimated beam making an equal but opposite angle with the axis of the optical system is used as the reference beam. Two exposures are made with light of two different wavelengths, λ_1 and λ_2. After processing, the hologram is replaced in its original position and illuminated with one of the wavelengths, say λ_2.

Interference fringes are seen due to the axial displacement of one image with respect to the other, which, from (2.3), is

$$\Delta z = z(\lambda_1 - \lambda_2)\big/\lambda_1. \tag{14.10}$$

Successive fringes correspond to increments of Δz of $\lambda_2/2$, or, when λ_1 and λ_2 are not very different, to increments of z given by the relation

$$\delta z = (\bar{\lambda})^2\big/2\Delta\lambda. \tag{14.11}$$

A fairly wide range of contour intervals can be obtained with pairs of lines from an Ar^+ laser. With a dye laser, the contour interval can be varied continuously.

14.3.2 Two-index contouring

In this technique [Tsuruta et al., 1967], the object is placed, as shown in fig. 14.5, in a cell with a plane glass window and viewed through a telecentric system. A beam splitter is used to illuminate the object with a collimated beam along the axis of the optical system. The hologram plate is located near the stop of the telecentric system.

Two holograms are recorded on the same plate with the cell filled with fluids having refractive indices n_1 and n_2, respectively. When the hologram is replaced

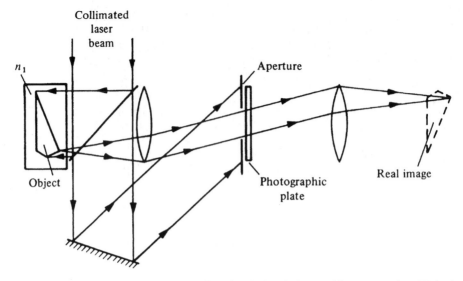

Fig. 14.5.   Optical system for two-refractive-index holographic contouring [Zelenka & Varner, 1969].

in the same position and illuminated once again by the same reference beam, one of the images is longitudinally displaced with respect to the other by an amount

$$\Delta z = (n_1 - n_2)z, \tag{14.12}$$

giving rise to fringes corresponding to increments of z given by the relation

$$\delta z = \lambda \big/ 2 \, |n_1 - n_2|. \tag{14.13}$$

The contouring interval can be varied from about 1 µm to 300 µm by using air and a liquid, or two liquids.

Typical contours obtained by both these methods using the same object are presented in fig. 14.6.

14.3.3 Changing the angle of illumination

A simple method of contouring is to make a double-exposure hologram with the object illuminated from two slightly different directions. With a point source, a small lateral displacement of the source between the two exposures produces contouring surfaces consisting of a set of hyperboloids of revolution, with the two positions of the source as their foci. Plane contouring surfaces can be obtained with collimated illumination.

To obtain contouring surfaces normal to the line of sight, the beam

Fig. 14.6. Depth contours obtained by (*a*) the two-wavelength method (contour interval 9.25 μm); (*b*) the two-index method (contour interval 11.8 μm) [Zelenka & Varner, 1969].

illuminating the object must also be normal to the line of sight. A convenient alternative is to use a sandwich hologram [Abramson, 1976]; the orientation of the contouring surfaces can then be controlled by tilting the sandwich.

References

Abramson, N. (1976). Sandwich hologram interferometry. 3. Contouring. *Applied Optics*, **15**, 200–5.

Ågren, C. H. & Stetson, K. A. (1972). Measuring the resonances of treble viol plates by hologram interferometry and designing an improved instrument. *Journal of the Acoustical Society of America*, **51**, 1971–83.

Archbold, E. & Ennos, A. E. (1968). Observation of surface vibration modes by stroboscopic hologram interferometry. *Nature*, **217**, 942–3.

Biedermann, K. & Molin, N.-E. (1970). Combining hypersensitization and *in situ* processing for time-average observation in real-time hologram interferometry. *Journal of Physics E: Scientific Instruments*, **3**, 669–80.

Bjelkhagen, H. (1974). Holographic time-average vibration study of a structure dynamic model of an airplane fin. *Optics & Laser Technology*, **6**, 117–23.

Chomat, M. & Miler, M. (1973). Application of holography to the analysis of mechanical vibration in electronic components. *TESLA Electronics*, **3**, 83–93.

Der, V. K., Holloway, D. C. & Fourney, W. L. (1973). Four-exposure holographic moiré technique. *Applied Optics*, **12**, 2552–4.

Haines, K. A. & Hildebrand, B. P. (1965). Contour generation by wavefront reconstruction. *Physics Letters*, **19**, 10–11.

Powell, R. L. & Stetson, K. A. (1965). Interferometric vibration analysis by wavefront reconstruction. *Journal of the Optical Society of America*, **55**, 1593–8.

Stetson, K. A. & Powell, R. L. (1965). Interferometric hologram evaluation and real-time vibration analysis of diffuse objects. *Journal of the Optical Society of America*, **55**, 1694–5.

Tsuruta, T., Shiotake, N., Tsujiuchi, J. & Matsuda, K. (1967). Holographic generation of contour map of diffusely reflecting surface by using immersion method. *Japanese Journal of Applied Physics*, **6**, 661–2.

Zelenka, J. S. & Varner, J. R. (1968). New method for generating depth contours holographically. *Applied Optics*, **7**, 2107–10.

Zelenka, J. S. & Varner, J. R. (1969). Multiple-index holographic contouring. *Applied Optics*, **8**, 1431–4.

Problems

14.1. If the time-average holograms of the guitar in fig. 14.2 had been recorded with the same optical system as that described for Problem 13.1, what would be the vibration amplitudes of the soundboard in the resonant modes at 195 and 292 Hz?

In these two modes, we have 7 dark fringes and 6 dark fringes, respectively, from the edge of the soundboard to the point vibrating with the largest amplitude. Since the edge is at rest, the amplitudes of vibration at these points correspond (see Section 14.2.2) to the sixth and seventh zeros of the function $J_0(\Omega)$, where $\Omega(x, y) = \mathbf{K} \cdot \mathbf{L}(x, y)$. Accordingly, we have

$$\Omega_{195} = 21.21, \tag{14.14}$$
$$\Omega_{195} = 18.07. \tag{14.15}$$

The corresponding values of the vibration amplitude are, therefore

$$\mathbf{L}_{195} = \frac{21.21}{16.94 \times 10^6} \text{ m}$$
$$= 1.25 \text{ μm}, \tag{14.16}$$

$$\mathbf{L}_{292} = \frac{18.07}{16.94 \times 10^6} \text{ m}$$
$$= 1.07 \text{ μm}. \tag{14.17}$$

Note that, for the mode at 292 Hz, a section of the soundboard between the two peaks is at rest; the displacements of the two peaks are therefore in opposite senses.

14.2. With the two-wavelength technique, what would be the contour interval obtained with the two spectral lines from an Ar^+ laser at $\lambda = 514$ nm and 488 nm?

From (14.11), the contour interval would be

$$\delta z \approx \frac{(0.501)^2}{2 \times 0.026} \text{ μm}$$
$$\approx 4.8 \text{ μm}. \tag{14.18}$$

15

Electronic techniques

Because the intensity in two-beam interference fringes varies sinusoidally with the phase difference, it is difficult to locate the fringe maxima or minima, in a photograph of the interference pattern, to better than a tenth of the fringe spacing. In addition, when the number of fringes is small and they are unequally spaced, errors are introduced by the need for nonlinear interpolation to determine the fractional fringe order at any point.

15.1 Computer-aided evaluation

One way to obtain higher accuracy is by using a CCD camera interfaced with a computer to sample and store the values of the intensity in the interference pattern at an array of points. These values can then be digitized and processed, using a number of techniques, to obtain the fractional fringe order at these points [Robinson & Reid, 1993]. Preprocessing is usually necessary to reduce speckle noise as well as to correct for local variations in image brightness.

15.1.1 Fourier-transform techniques

An additional tilt introduced in one of the beams (say, along the x direction) generates background fringes corresponding to a spatial carrier frequency. These fringes are modulated by the additional phase difference between the beams due to the changes in the object [Takeda, Ina & Kobayashi, 1982]. If the spatial carrier frequency is sufficiently high, the Fourier transform of the intensity distribution in the interference pattern can be processed to obtain the phase difference.

The Fourier-transform technique is most useful in studies of phase objects, where the sensitivity vector does not vary over the field and it is possible to produce straight, equally spaced carrier fringes [Bone, Bachor & Sandeman, 1986].

15.1.2 Phase unwrapping

Since the values of the phase difference obtained at any point are known only to modulo 2π, it is necessary to determine the number of 2π steps to be added to these raw values. This process is known as phase unwrapping.

Phase unwrapping requires a knowledge of the sign as well as the magnitude of the raw phase. It is then possible, by choosing a starting point at which the phase difference is known to be zero, and checking the values of the raw phase at adjacent pixels along a line, to decide whether to add or subtract 2π when crossing successive fringes. This procedure is extended to two dimensions by using pixels along the first line as new starting points [Bone, 1991].

15.2 Heterodyne interferometry

Very accurate measurements of the phase difference at any point can be made by heterodyne holographic interferometry.

In this technique [Dändliker, 1980], two holograms are recorded of the object, at successive stages of loading, using the optical system shown in fig. 15.1. The two holograms are recorded on the same plate, with different reference beams having the same frequency as the object beam, but with an angular separation.

At the reconstruction stage, a small frequency difference is introduced between the two reference beams illuminating the hologram by means of a rotating grating or two acousto-optic modulators operated at slightly different frequencies. When the two reconstructed waves are superposed at a photodetector, the output is given by the relation

$$I(x, y, t) = |a_1(x, y)|^2 + |a_2(x, y)|^2 + 2|a_1(x, y)||a_2(x, y)| \times \cos[2\pi(\nu_1 - \nu_2)t - \Delta\phi(x, y)], \tag{15.1}$$

where ν_1 and ν_2 are the frequencies of the two beams illuminating the hologram, $a_1(x, y)$ and $a_2(x, y)$ are the amplitudes of the reconstructed waves and $\Delta\phi(x, y) = \phi_1(x, y) - \phi_2(x, y)$, where $\phi_1(x, y)$ and $\phi_2(x, y)$ are the phases of the reconstructed waves. The output from the photodetector is modulated at the beat frequency $\nu_1 - \nu_2$, and the phase $\Delta\phi(x, y)$ of the modulation corresponds to the phase difference between the reconstructed waves. This phase can be measured electronically by comparison with a reference signal from a second

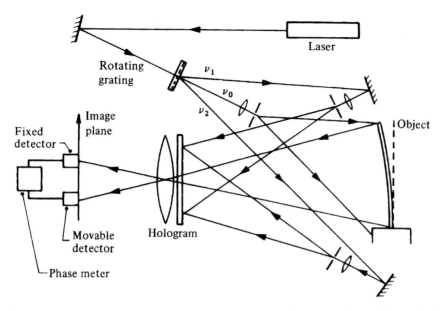

Fig. 15.1.   System for heterodyne holographic interferometry [Dändliker, 1980].

fixed detector, as the first detector is moved in steps across the interference field.

This technique can measure phase differences with a high degree of accuracy (better than $2\pi/500$), but, since it involves point-by-point measurements, it is slow and requires a very stable environment.

15.3 Phase-shifting interferometry

Phase-shifting holographic interferometry permits accurate measurements on real-time fringes at a very large number of points in a very short time [Hariharan, Oreb & Brown, 1982].

In this technique, as shown in fig. 15.2, an image of the real-time fringes is formed on a CCD array. During a single scan of the array, the values of the intensity at each of the pixels are read out, digitized and stored in the memory of a computer. Between successive scans of the array, the phase of the reference beam is shifted, relative to that of the object beam, by means of a mirror mounted on a piezoelectric translator (PZT) to which appropriate voltages are applied by an amplifier controlled, through a digital-to-analog converter, by the computer.

At least three sets of intensity data are required to evaluate the phase distribution in the interference pattern, but the most common algorithm for phase

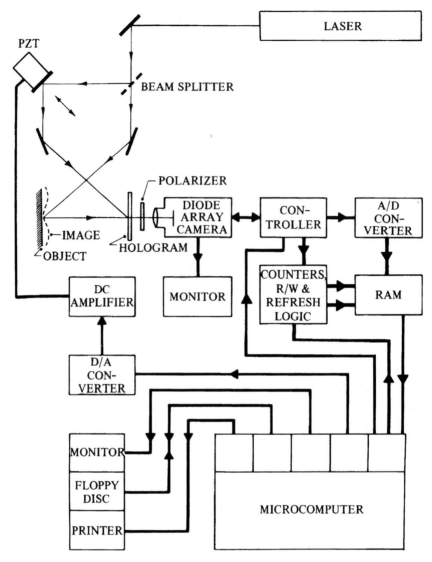

Fig. 15.2.   System for phase-shifting holographic interferometry [Hariharan, Oreb & Brown, 1982].

calculations [Creath, 1988] uses four frames with phase shifts of 0°, 90°, 180° and 270°. The intensity values at any point can then be written as

$$I_0 = I_{Av}[1 + V \cos \phi], \qquad (15.2)$$

$$I_{90} = I_{Av}[1 - V \sin \phi], \qquad (15.3)$$

$$I_{180} = I_{Av}[1 - V \cos \phi], \qquad (15.4)$$

$$I_{270} = I_{Av}[1 + V \sin \phi], \qquad (15.5)$$

where I_{Av} is the average intensity, $\mathbf{V}$ is the visibility of the fringes and ϕ is the original phase difference between the beams, and the original phase difference between the beams is given by the relation

$$\phi = \arctan\left(\frac{I_{270} - I_{90}}{I_0 - I_{180}}\right). \tag{15.6}$$

Phase-shifting interferometry does not offer as high precision as heterodyne interferometry but has the advantage that measurements are made simultaneously over the entire array of points, so that the results are less sensitive to environmental effects. Typically, the intensity data for a 512×512 array of points can be acquired in less than a second and processed to obtain values of the phase difference, with an accuracy of $2\pi/200$, in a few seconds.

15.3.1 Vector displacements

Phase-shifting has been applied to the measurement of vector displacements and strains. The optical system used permits four holograms to be recorded in quick succession on a photothermoplastic material with the object illuminated from two different directions in the horizontal and vertical planes [Hariharan, Oreb & Brown, 1983].

By combining the phase data from the real-time fringes obtained with these four holograms with data on the shape of the object, it is possible to evaluate the in-plane and normal components of the surface displacement at each point. Differentiation of the displacements then gives the strains [Zarrabi, Oreb & Hariharan, 1990].

15.3.2 Contouring

Phase-shifting can be used with two-wavelength holography [Wyant, Oreb & Hariharan, 1984], as well as with the two-refractive-index technique [Hariharan & Oreb, 1984], for contouring surfaces. Figure 15.3 shows a contour map and an isometric plot of the wear pattern on a surface obtained with the two-refractive-index technique. The contour interval was 200 μm, and readings could be repeated to ± 1 μm.

15.3.3 Vibration analysis

Stroboscopic holographic interferometry yields fringes with a cosinusoidal intensity distribution, so that accurate measurements of the surface displacement can be made by phase-shifting techniques. However, when observing the

(a)   (b)

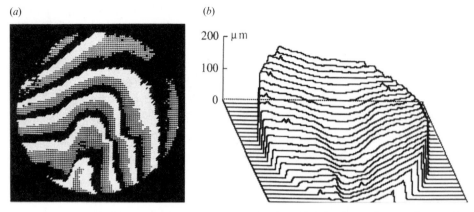

Fig. 15.3.   (*a*) Contour map and (*b*) isometric plot of a wear mark on a flat surface obtained by phase-shifting holographic interferometry using the two-refractive-index technique [Hariharan & Oreb, 1984].

(a)   (b)

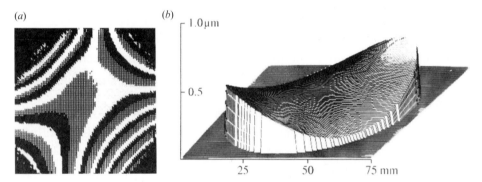

Fig. 15.4.   (*a*) Contour plot of the surface displacements of a square metal plate vibrating at 231 Hz, obtained by phase-shifting holographic interferometry with stroboscopic illumination, and (*b*) a three-dimensional plot of the displacements over the center of the plate [Hariharan & Oreb, 1986].

real-time fringes, the pulse width has to be reduced to a small fraction of the period of the vibration, resulting in a serious drop in the brightness of the image. This problem can be overcome by using stroboscopic illumination to record a hologram of the vibrating object and making measurements on the fringes obtained with the stationary object and continuous illumination [Hariharan & Oreb, 1986].

Figure 15.4(*a*) is a contour plot of the surface displacements of a square metal plate vibrating at 231 Hz obtained by phase-shifting holographic interferometry with stroboscopic illumination, while fig. 15.4(*b*) is a three-dimensional plot of the displacements over the center of the plate.

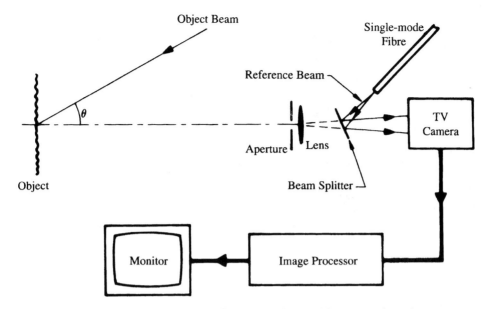

Fig. 15.5. Schematic of a system for electronic speckle-pattern interferometry.

Phase-shifting has also been used for vibration analysis with time-averaged fringes [Nakadate, 1986].

15.4 Electronic holographic interferometry

Electronic holographic interferometry can be regarded as having evolved from electronic speckle-pattern interferometry (ESPI) [Butters & Leendertz, 1971; Macovski, Ramsay & Schaefer, 1971; for more details on the development of ESPI, see Jones & Wykes, 1989].

A typical system used for ESPI is shown in fig. 15.5. The object is imaged on the target of a television camera along with a coaxial reference beam. The resulting image hologram has a coarse speckle structure which can be resolved by the television camera. Any change in the shape of the object results in a change in the intensity distribution in the speckles in the corresponding part of the image.

To measure displacements of the object, an image of the object in its original state is stored and subtracted from the signal from the television camera. Regions in which the speckle pattern has not changed, corresponding to the condition

$$\mathbf{K} \cdot \mathbf{L}(x, \, y) = 2m\pi, \tag{15.7}$$

where m is an integer, then appear dark, while regions where the speckle pattern has changed (see Section 2.5) appear covered with bright speckles.

The application of phase-shifting techniques made possible direct measurements of surface displacements [Creath, 1985; Robinson & Williams, 1986]. Each speckle, as seen by the camera, can be regarded as an individual interference pattern, and the phase difference between the beams at this point can be measured before and after the object is deformed. Even though the initial phase differences in neighboring speckles may be very different, the changes in the phase differences will be the same for the same surface displacement. Accordingly, the result of subtracting the second set of phase values from the first is a contour map of the changes in the shape of the object.

In electronic holographic interferometry, the phase change is calculated directly from two sets of four frames of intensity data acquired with phase increments of $\pi/2$. One set is recorded before, and the other after, the object is stressed. The change in the phase difference at any point can then be obtained from the relation

$$\Delta\phi = \arctan\left[\frac{\sin(\phi - \phi')}{\cos(\phi - \phi')}\right]$$

$$= \arctan\left[\frac{\sin\phi\,\cos\phi' - \cos\phi\,\sin\phi'}{\cos\phi\,\cos\phi' + \sin\phi\,\sin\phi'}\right]$$

$$= \arctan\left[\frac{(I_4 - I_2)(I'_1 - I'_3) - (I_1 - I_3)(I'_4 - I'_2)}{(I_1 - I_3)(I'_1 - I'_3) - (I_4 - I_2)(I'_4 - I'_2)}\right]. \qquad (15.8)$$

A problem with electronic holographic interferometry is that the phase data are noisy, due to the coarse speckle structure of the object beam. The noise can be reduced by averaging several sets of phase data obtained with slightly different directions of illumination, so that the speckle patterns in the images are uncorrelated.

15.4.1 Vibration analysis

Values of the vibration amplitude can also be obtained from time-average holograms by phase-shifting [Stetson & Brohinsky, 1988].

Three sets of four intensity measurements are used. The first set of measurements is made with the object vibrating and additional phase shifts of 0, $\pi/2$, π and $3\pi/2$ introduced in the reference beam. The second and third sets of measurements are made with a sinusoidal phase modulation introduced in the reference beam at the vibration frequency. The phase of this modulation is offset from the phase of the vibration by $+\pi/3$ in one case and $-\pi/3$ in the other case. The twelve sets of data can then be processed to obtain the vibration amplitude at each point.

References

Bone, D. J. (1991). Fourier fringe analysis: the two-dimensional phase unwrapping problem. *Applied Optics*, **30**, 3627–32.

Bone, D. J., Bachor, H.-A. & Sandeman, R. J. (1986). Fringe-pattern analysis using a 2-D Fourier transform. *Applied Optics*, **25**, 1653–60.

Butters, J. N. & Leendertz, J. A. (1971). A double-exposure technique for speckle-pattern interferometry. *Journal of Physics E: Scientific Instruments*, **4**, 277–9.

Creath, K. (1985). Phase-shifting speckle interferometry. *Applied Optics*, **24**, 3053–8.

Creath, K. (1988). Phase-measurement interferometry techniques. In *Progress in Optics*, vol. 26, ed. E. Wolf, pp. 349–93. Amsterdam: North-Holland.

Dändliker, R. (1980). Heterodyne holographic interferometry. In *Progress in Optics*, vol. 17, ed. E. Wolf, pp. 1–84. Amsterdam: North-Holland.

Hariharan, P., Oreb, B. F. & Brown, N. (1982). A digital phase-measurement system for real-time holographic interferometry. *Optics Communications*, **41**, 393–6.

Hariharan, P., Oreb, B. F. & Brown, N. (1983). Real-time holographic interferometry: a microcomputer system for the measurement of vector displacements. *Applied Optics*, **22**, 876–80.

Hariharan, P. & Oreb, B. F. (1984). Two-index holographic contouring: application of digital techniques. *Optics Communications*, **51**, 142–4.

Hariharan, P. & Oreb, B. F. (1986). Stroboscopic holographic interferometry: application of digital techniques. *Optics Communications*, **59**, 83–6.

Jones, R. & Wykes, C. (1989). *Holographic & Speckle Interferometry*. Cambridge: Cambridge University Press.

Macovski, A., Ramsay, S. D. & Schaefer, L. F. (1971). Time-lapse interferometry and contouring using television systems. *Applied Optics*, **10**, 2722–7.

Nakadate, S. (1986). Vibration measurement using phase-shifting time-average holographic interferometry. *Applied Optics*, **25**, 4155–61.

Robinson, D. W. & Reid, G. T., eds. (1993). *Interferogram Analysis: Digital Processing Techniques for Fringe Pattern Measurement*. London: IOP.

Robinson, D. W. & Williams, D. C. (1986). Digital phase stepping speckle interferometry. *Optics Communications*, **57**, 26–30.

Stetson, K. A. & Brohinsky, W. R. (1988). Fringe-shifting techniques for numerical analysis of time-average holograms of vibrating objects. *Journal of the Optical Society of America. A.* **5**, 1472–6.

Takeda, M., Ina, H. & Kobayashi, S. (1982). Fourier-transform method of fringe-pattern analysis for computer-based topography and interferometry. *Journal of the Optical Society of America.* **72**, 156–60.

Wyant, J. C., Oreb, B. F. & Hariharan, P. (1984). Testing aspherics using two-wavelength holography: application of digital electronic techniques. *Applied Optics*, **23**, 4020–3.

Zarrabi, K., Oreb, B. F. & Hariharan, P. (1990). Laser holographic interferometry and finite element analysis: a comparative assessment with reference to pressure vessels. *Non-destructive Testing Australia*, **27**, 64–8.

Appendix A

Interference and coherence

A.1 Interference

The time-varying electric field E at any point due to a linearly polarized monochromatic light wave propagating in a vacuum in the z direction can be represented by the relation

$$E = a \cos[2\pi\nu(t - z/c)],\qquad\text{(A.1)}$$

where a is the amplitude, ν the frequency and c the speed of propagation of the light wave. Equation (A.1) can be written in the form

$$E = \mathrm{Re}\{a \exp[\mathrm{i}2\pi\nu(t - z/c)]\},$$

$$= \mathrm{Re}\{a \exp(-\mathrm{i}\phi) \exp(\mathrm{i}2\pi\nu t)\},\qquad\text{(A.2)}$$

where $\mathrm{Re}\{\ldots\}$ represents the real part of the expression within the braces, $\mathrm{i} = (-1)^{1/2}$, and $\phi = 2\pi\nu z/c = 2\pi z/\lambda$.

If we assume that all operations on E are linear, we can use the complex representation

$$E = a \exp(-\mathrm{i}\phi) \exp(\mathrm{i}2\pi\nu t),$$

$$= A \exp(\mathrm{i}2\pi\nu t),\qquad\text{(A.3)}$$

where $A = a \exp(-\mathrm{i}\phi)$ is known as the complex amplitude. Multiplication of the complex amplitude by its complex conjugate yields the optical intensity

$$I = AA^* = |A|^2.\qquad\text{(A.4)}$$

Since the complex amplitude at any point due to a number of waves of the same frequency is the sum of the complex amplitudes of the individual waves

$$A = A_1 + A_2 + \cdots,\qquad\text{(A.5)}$$

the intensity due to the interference of two waves is

$$I = |A_1 + A_2|^2,$$

$$= |A_1|^2 + |A_2|^2 + A_1 A_2^* + A_1^* A_2,$$

$$= I_1 + I_2 + 2(I_1 I_2)^{1/2} \cos(\phi_1 - \phi_2).\qquad\text{(A.6)}$$

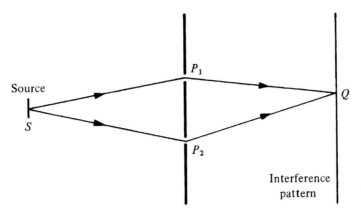

Fig. A.1.   Evaluation of the degree of coherence.

The visibility of the interference fringes is defined as

$$V = \frac{I_{max} - I_{min}}{I_{max} + I_{min}} = \frac{2(I_1 I_2)^{1/2}}{I_1 + I_2}.$$ (A.7)

Equations (A.6) and (A.7) assume that the electric vectors of the two waves are parallel. If the electric vectors make an angle θ with each other, the visibility of the interference fringes is only

$$V_\theta = V \cos \theta,$$ (A.8)

which drops to zero when $\theta = \pi/2$.

A.2 Coherence

We have assumed in the previous section that the light waves are derived from a single point source emitting an infinitely long, monochromatic wave train, in which case we say that the fields due to the two light waves are perfectly coherent. In reality, all wave fields are only partially coherent.

We can represent the time-varying electric field at any point due to a linearly polarized, quasi-monochromatic light wave from a source of finite size by the analytic signal [Born & Wolf, 1999]

$$V(t) = \int_0^\infty a(v) \exp[-i\phi(v) \exp(i2\pi vt)]dv,$$ (A.9)

where $a(v)$ is the amplitude and $\phi(v)$ is the phase of a component with frequency v. To evaluate the degree of coherence of the fields at two points illuminated by such a source, we consider the optical system shown in fig. A.1. In this arrangement, if $V_1(t)$ and $V_2(t)$ are the analytic signals corresponding to the electric fields at P_1 and P_2, the complex degree of coherence $\gamma_{12}(\tau)$ of the fields, for a time delay τ (the difference in the transit times from S to P_1 and P_2), is defined as the normalized correlation of $V_1(t)$ and $V_2(t)$ (see Appendix B)

$$\gamma_{12}(\tau) = \frac{\langle V_1(t+\tau)V_2^*(t)\rangle}{[\langle V_1(t)V_1^*(t)\rangle\langle V_2(t)V_2^*(t)\rangle]^{1/2}}.$$ (A.10)

The physical significance of (A.10) can be understood if the light waves are allowed to emerge through pinholes at P_1 and P_2, so that they form an interference pattern on a screen. P_1 and P_2 can then be considered as two secondary sources, so that, from (A.6), the intensity at Q is

$$I = I_1 + I_2 + \langle V_1(t+\tau)V_2^*(t) + V_1^*(t+\tau)V_2(t)\rangle,$$

$$= I_1 + I_2 + 2\text{Re}[\langle V_1(t+\tau)V_2^*(t)\rangle], \tag{A.11}$$

where I_1 and I_2 are the intensities at Q when P_1 and P_2 act separately, and τ is the difference in the transit times for the paths P_1Q and P_2Q. We can rewrite this relation, from (A.10), as

$$I = I_1 + I_2 + 2(I_1 I_2)^{1/2} \, \text{Re}[\gamma_{12}(\tau)]$$

$$= I_1 + I_2 + 2(I_1 I_2)^{1/2} |\gamma_{12}(\tau)| \cos \phi_{12}(\tau), \tag{A.12}$$

where $\phi_{12}(\tau)$ is the phase of $\gamma_{12}(\tau)$.

Interference fringes are produced by the variations in $\cos \phi_{12}(\tau)$ across the screen. If $I_1 = I_2$, the visibility of these fringes is, from (A.7),

$$\mathbf{V} = |\gamma_{12}(\tau)|. \tag{A.13}$$

The coherence of the field produced by any light source can be studied from two aspects.

A.2.1 Spatial coherence

When the difference in the optical paths from the source to P_1 and P_2 is negligibly small, so that $\tau \approx 0$, effects due to the spectral bandwidth of the source can be neglected, and we are essentially concerned with what is termed the spatial coherence of the field. A special case of interest is when the dimensions of the source and the separation of P_1 and P_2 are extremely small compared to the distances between the source and P_1 and P_2. In this case, the complex degree of coherence is given by the normalized Fourier transform (see Appendix B) of the intensity distribution over the source.

A.2.2 Temporal coherence

If the source is effectively a point source, but radiates over a range of wavelengths, we are concerned with the temporal coherence of the field. In this case, the complex degree of coherence depends only on τ, the difference in the transit times. Equation (A.10) can then be transformed and written as

$$\gamma(\tau) = \frac{\mathcal{F}\{S(\nu)\}}{\int_{-\infty}^{\infty} S(\nu)d\nu}, \tag{A.14}$$

where $S(\nu)$ is the frequency spectrum of the radiation. In this case, it follows, from (A.12) and (A.13), that the degree of temporal coherence can be obtained from the visibility of the interference fringes as the difference in the lengths of the optical paths from the source is varied.

This argument leads us to the concepts of the coherence time and the coherence length of the radiation. It can be shown that, with radiation having a bandwidth $\Delta\nu$, the visibility of the interference fringes drops to zero for a difference in the transit times

$$\Delta\tau \approx 1/\Delta\nu. \tag{A.15}$$

This time $\Delta\tau$ is called the coherence time of the radiation; the corresponding value of the coherence length is

$$\Delta l \approx c\Delta\tau \approx \lambda_0^2/\Delta\lambda, \tag{A.16}$$

where λ_0 is the mean wavelength and $\Delta\lambda$ the bandwidth of the radiation. To obtain interference fringes with good visibility, the optical path difference must be small compared to the coherence length of the radiation.

References

Born, M. & Wolf, E. (1999). *Principles of Optics*. Cambridge: Cambridge University Press.

Appendix B

Fourier transforms

B.1 Two-dimensional transforms

A function of two orthogonal spatial coordinates can be expressed, by means of a two-dimensional Fourier transform, as a function of two orthogonal spatial frequencies [Goodman, 1996].

The two-dimensional Fourier transform of $g(x, y)$ is defined as

$$\mathcal{F}\{g(x, y)\} = \int_{-\infty}^{\infty} \int_{-\infty}^{\infty} g(x, y) \exp[-\mathrm{i}2\pi(\xi x + \eta y)]\mathrm{d}x\mathrm{d}y,$$

$$= G(\xi, \eta), \qquad (B.1)$$

where ξ and η are spatial frequencies. Similarly, the inverse Fourier transform of $G(\xi, \eta)$ is defined as

$$\mathcal{F}^{-1}\{G(\xi, \eta)\} = \int_{-\infty}^{\infty} \int_{-\infty}^{\infty} G(\xi, \eta) \exp[\mathrm{i}2\pi(\xi x + \eta y)]\mathrm{d}\xi\mathrm{d}\eta,$$

$$= g(x, y). \qquad (B.2)$$

These relationships can be written symbolically as

$$g(x, y) \leftrightarrow G(\xi, \eta). \qquad (B.3)$$

The Fourier transform effectively decomposes a complex wavefront into component plane waves whose propagation can then be analyzed as discussed in Appendix C.

B.2 Convolution

The convolution of a pair of two-dimensional functions $g(x, y)$ and $h(x, y)$ is

$$f(x, y) = \int_{-\infty}^{\infty} \int_{-\infty}^{\infty} g(u, v)h(x - u, y - v)\mathrm{d}u\mathrm{d}v, \qquad (B.4)$$

which can be written as

$$f(x, y) = g(x, y) * h(x, y), \qquad (B.5)$$

where the symbol $*$ denotes the convolution operation.

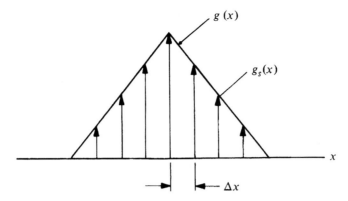

Fig. B.1.   The function $g_s(x)$ consists of an array of delta functions obtained by sampling the function $g(x)$ at intervals of Δx.

By definition, convolution of a function with the Dirac delta function $\delta(x, y)$ yields the original function.

B.3 Correlation

The cross-correlation of two functions, $g(x, y)$ and $h(x, y)$ is

$$c(x, y) = \int_{-\infty}^{\infty} \int_{-\infty}^{\infty} g^*(u, v)h(x+u, y+v) du\, dv, \tag{B.6}$$

where $g^*(u, v)$ is the complex conjugate of $g(u, v)$. This equation can be written as

$$c(x, y) = g(x, y) \star h(x, y), \tag{B.7}$$

where the symbol $\star$ denotes the correlation operation.
    It follows that the autocorrelation of a function $g(x, y)$ is

$$a(x, y) = g(x, y) \star g(x, y). \tag{B.8}$$

B.4 Sampled functions

The production of computer-generated holograms involves computation of the Fourier transform $G_s(\xi, \eta)$ of a sampled function $g_s(x, y)$, which is obtained by sampling $g(x, y)$, the wave to be reconstructed, at intervals $(\Delta x, \Delta y)$. If, for simplicity, we consider the one-dimensional case, as shown in fig. B.1, we can write

$$g_s(x) = g(x) \sum_{m=-\infty}^{\infty} \delta(x - m\Delta x), \tag{B.9}$$

and

$$G_s(\xi) = (1/\Delta x) \sum_{m=-\infty}^{\infty} G[\xi - (m/\Delta x)], \tag{B.10}$$

where $G(\xi) \leftrightarrow g(x)$.

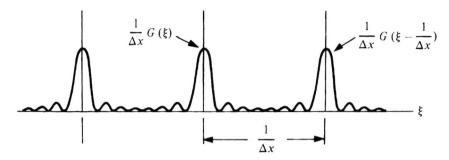

Fig. B.2.   The Fourier transform $G_s(\xi)$ of the sampled function $g_s(x)$ consists of a regular series of repetitions of $G(\xi)$, the Fourier transform of $g(x)$, shifted in frequency by successive intervals $\Delta\xi = 1/\Delta x$.

As shown in fig. B.2, the Fourier transform of the sampled function $g_s(x)$ is a regular series of repetitions of the Fourier transform of the original function $g(x)$, shifted in frequency by successive intervals $\Delta\xi = (1/\Delta x)$.

If the interval Δx at which $g(x)$ is sampled is small enough to avoid aliasing (overlapping of the shifted Fourier transforms),

$$\Delta x \le 1/\xi_{max}, \tag{B.11}$$

it is possible to recover $G(\xi)$ from $G_s(\xi)$.

References

Goodman, J. W. (1996). *Introduction to Fourier Optics*. New York: McGraw-Hill.

Appendix C

Wave propagation

We consider, as shown in fig. C.1, a plane wave with an amplitude a incident normally on an object having an amplitude transmittance $\mathbf{t}(x_1, y_1)$ located in the plane $z = 0$.

The complex amplitude $a(x, y, z)$ at a point $P(x, y, z)$ located at a distance z is then given by the Fresnel–Kirchhoff integral [Born & Wolf, 1999] which, if z is much larger than $(x - x_1)$ and $(y - y_1)$, can be simplified and written as

$$a(x, y, z) = (ia/\lambda z) \int_{-\infty}^{\infty} \int_{-\infty}^{\infty} \mathbf{t}(x_1, y_1)$$

$$\times \exp\{(-i\pi/\lambda z)[(x - x_1)^2 + (y - y_1)^2]\} dx_1 dy_1 \qquad (C.1)$$

$$= (ia/\lambda z) \int_{-\infty}^{\infty} \int_{-\infty}^{\infty} \mathbf{t}(x_1, y_1) \exp[(-i\pi/\lambda z)(x^2 + y^2)]$$

$$\times \exp[(-i\pi/\lambda z)(x_1^2 + y_1^2)]$$

$$\times \exp\{i2\pi[x_1(x/\lambda z) + y_1(y/\lambda z)]\} dx_1 dy_1. \qquad (C.2)$$

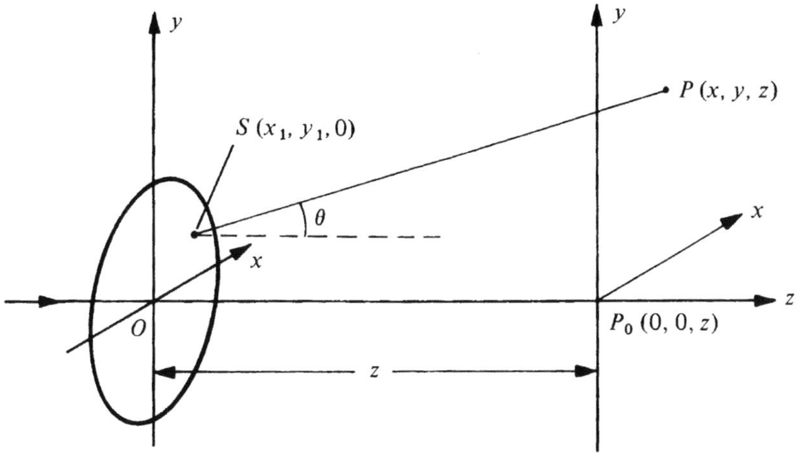

Fig. C.1.   Coordinate system used to evaluate the Fresnel–Kirchhoff integral.

151

Since the first exponential factor in (C.2) is independent of x_1 and y_1, it can be taken outside the integral sign. In addition, if the distance to the plane of observation is large compared to the dimensions of the object, so that

$$z \gg (x_1^2 + y_1^2)/\lambda, \tag{C.3}$$

(the far-field condition), the second exponential factor is close to unity. If then we set

$$\xi = x/\lambda z, \tag{C.4}$$

$$\eta = y/\lambda z, \tag{C.5}$$

where ξ and η denote spatial frequencies, (C.2) becomes

$$a(x, y, z) = (ia/\lambda z) \exp[(-i\pi/\lambda z)(x^2 + y^2)]$$

$$\times \int_{-\infty}^{\infty} \int_{-\infty}^{\infty} \mathbf{t}(x_1, y_1) \exp[i2\pi(\xi x_1 + \eta y_1)]dx_1 dy_1$$

$$= (ia/\lambda z) \exp[(-i\pi/\lambda z)(x^2 + y^2)]\mathbf{T}(\xi, \eta), \tag{C.6}$$

where

$$\mathbf{t}(x_1, y_1) \leftrightarrow \mathbf{T}(\xi, \eta). \tag{C.7}$$

It follows that the complex amplitude in the plane of observation is given by the Fourier transform of the amplitude transmittance of the object, multiplied by a spherical phase factor.

References

Born, M. & Wolf, E. (1999). *Principles of Optics*. Cambridge: Cambridge University Press.

Appendix D

Speckle

When a diffusely reflecting surface is illuminated with a laser, each of the microscopic elements on the surface produces a diffracted wave. Since the optical paths to neighboring elements exhibit random differences which may amount to several wavelengths, the intensity in the far field exhibits violent local fluctuations. The general appearance of the resulting speckle pattern (see fig. D.1) is almost independent of the characteristics of the surface, but the scale of the granularity depends on the size of the diffusing surface and the distance at which the pattern is observed.

D.1 First-order statistics

The complex amplitude at any point in the far field of a diffusing surface illuminated by a laser is obtained by summing the complex amplitudes of the diffracted waves from the individual elements on the surface. For polarized light, we have

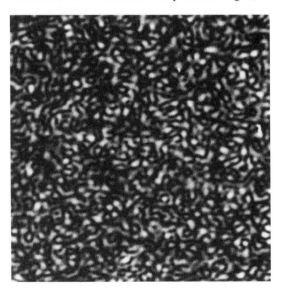

Fig. D.1.   Speckle pattern observed when a diffusing surface is illuminated with a laser.

153

$$a \exp(-i\phi) = \sum_{n=1}^{n=n} a_n \exp(-i\phi_n). \tag{D.1}$$

If we assume that the moduli of the individual complex amplitudes are equal, while their phases, after subtracting integral multiples of 2π, are uniformly distributed over the range from 0 to 2π, this reduces to the well-known random-walk problem. The joint probability density function of the real and imaginary parts of the complex amplitude is then [Goodman, 1975]

$$p_{r,i}(a_{r,i}) = \frac{1}{2\pi\sigma^2} \exp\left[\frac{-(a_r^2 + a_i^2)}{2\sigma^2}\right], \tag{D.2}$$

where σ^2 is a constant. The most common value of the modulus is zero, while the phase has a uniform circular distribution.

It can then be shown that the probability density function of the intensity is the negative exponential distribution

$$p(I) = \frac{1}{2\sigma^2} \exp\left(\frac{-I}{2\sigma^2}\right). \tag{D.3}$$

It follows that the mean value of the intensity is

$$\langle I \rangle = 2\sigma^2, \tag{D.4}$$

and its second moment is

$$\langle I^2 \rangle = 2\langle I \rangle^2. \tag{D.5}$$

Accordingly, the variance of the intensity is

$$\sigma_I^2 = \langle I^2 \rangle - \langle I \rangle^2,$$
$$= \langle I \rangle^2, \tag{D.6}$$

and the contrast of the speckle pattern, defined by the relation,

$$c = \frac{\sigma_I}{\langle I \rangle}, \tag{D.7}$$

is equal to unity.

D.2 Second-order statistics

The autocorrelation function and the power spectral density of the intensity distribution can be evaluated by making use of the fact that the complex amplitude at the scattering surface and the complex amplitude in the plane of observation are related by the Fresnel–Kirchhof integral.

For a square scattering surface whose edges have a length L, it can be shown that the average dimensions of a single speckle in a speckle pattern observed at a distance z are

$$\delta x = \delta y = \lambda z / L, \tag{D.8}$$

while the power spectral density of the intensity distribution, apart from a delta function at zero spatial frequency, is the normalized autocorrelation function of the intensity distribution over the scattering surface.

D.3 Image speckle

We can regard the entrance pupil of the imaging system as being illuminated by the primary speckle pattern. The intensity distribution in the image can then be obtained by treating the exit pupil as a diffuser.

It follows that with a lens having a circular aperture of radius ρ, the average size of the speckles in the image is, from (D.8),

$$\delta x = \delta y = 0.61 \, \lambda f/\rho, \qquad (D.9)$$

where f is the focal length of the lens.

References

Goodman, J. W. (1975). Statistical properties of laser speckle patterns. In *Laser Speckle & Related Phenomena*, Topics in Applied Physics, vol. 9, ed. J. C. Dainty, pp. 9–75. Berlin: Springer-Verlag.

Bibliography

Abramson, N. (1981). *The Making & Evaluation of Holograms*. London: Academic Press.

Beiser, L. (1988). *Holographic Scanning*. New York: Wiley.

Bjelkhagen, H. I. (1993). *Silver Halide Materials for Holography & Their Processing*. Berlin: Springer-Verlag.

Collings, N. (1988). *Optical Pattern Recognition: Using Holographic Techniques*. Reading: Addison-Wesley.

Hariharan, P. (1996). *Optical Holography*. Cambridge: Cambridge University Press.

Jones, R. & Wykes, C. (1989). *Holographic & Speckle Interferometry*. Cambridge: Cambridge University Press.

Okoshi, T. (1976). *Three Dimensional Imaging Techniques*. New York: Academic Press.

Ostrovsky, Yu. I. (1991). *Holographic Interferometry in Experimental Mechanics*. Berlin: Springer-Verlag.

Rastogi, P. K., ed. (1994). *Holographic Interferometry*. Berlin: Springer-Verlag.

Robillard, J. & Caulfield, H. J., eds. (1990). *Industrial Applications of Holography*. New York: Oxford University Press.

Robinson, D. W. & Reid, G. T., eds. (1993). *Interferogram Analysis: Digital Processing Techniques for Fringe Pattern Measurement*. London: IOP.

Saxby, G. (1991). *Manual of Practical Holography*. London: Focal Press.

Smith, H. M., ed. (1977). *Holographic Recording Materials*. Berlin: Springer-Verlag.

Solymar, L. & Cooke, D. J. (1981). *Volume Holography & Volume Gratings*. New York: Academic Press.

Syms, R. R. A. (1990). *Practical Volume Holography*. Oxford: Oxford University Press.

Vest, C. M. (1979). *Holographic Interferometry*. New York: John Wiley.

Vikram, C. (1992). *Particle Field Holography*. Cambridge: Cambridge University Press.

von Bally, G., ed. (1979). *Holography in Medicine & Biology*. Berlin: Springer-Verlag.

Index